Faruk Hadžiselimović

Cryptorchidism

Ultrastructure of Normal and Cryptorchid Testis Development

With 43 Figures

Springer-Verlag Berlin · Heidelberg · New York 1977

The Advances publish reviews and critical articles covering the entire field of normal anatomy (cytology, histology, cyto- and histochemistry, electron microscopy, macroscopy, experimental morphology and embryology and comparative anatomy). Papers dealing with anthropology and clinical morphology will also be accepted with the aim of encouraging co-operation between anatomy and related disciplines.

Papers, which may be in English, French or German, are normally commissioned, but original papers and communications may be submitted and will be considered so long as they deal with a subject comprehensively and meet the requirements of the "Advances".

For speed of publication and breadth of distribution, this journal appears in single issues which can be purchased separately; 6 issues constitute one volume.

It is a fundamental condition that submitted manuscripts have not been, and will not simultaneously be submitted or published elsewhere. With the acceptance of a manuscript for publication, the publisher acquire full and exclusive copyright for all languages and countries. 25 copies of each paper are supplied free of charge.

Die Ergebnisse dienen der Veröffentlichung zusammenfassender und kritischer Artikel aus dem Gesamtgebiet der normalen Anatomie (Cytologie, Histologie, Cyto- und Histochemie, Elektronenmikroskopie, Makroskopie, experimentelle Morphologie und Embryologie und vergleichende Anatomie). Aufgenommen werden ferner Arbeiten anthropologischen und morphologisch-klinischen Inhalts, mit dem Ziel, die Zusammenarbeit zwischen Anatomie und Nachbardisziplinen zu fördern.

Zur Veröffentlichung gelangen in erster Linie angeforderte Manuskripte, jedoch werden auch eingesandte Arbeiten und Originalmitteilungen berücksichtigt, sofern sie ein Gebiet umfassend abhandeln und den Anforderungen der „Ergebnisse" genügen. Die Veröffentlichungen erfolgen in englischer, deutscher und französischer Sprache.

Die Arbeiten erscheinen im Interesse einer raschen Veröffentlichung und einer weiten Verbreitung als einzeln berechnete Hefte; je 6 Hefte bilden einen Band.

Grundsätzlich dürfen nur Arbeiten eingesandt werden, die nicht gleichzeitig an anderer Stelle zur Veröffentlichung eingereicht oder bereits veröffentlicht worden sind. Der Autor verpflichtet sich, seinen Beitrag auch nachträglich nicht an anderer Stelle zu publizieren. Die Mitarbeiter erhalten von ihren Arbeiten zusammen 25 Freiexemplare.

Les résultats publient des sommaires et des articles critiques concernant l'ensemble du domaine de l'anatomie normale (cytologie, histologie, cyto- et histochimie, microscopie électronique, macroscopie, morphologie expérimentale, embryologie et anatomie comparée). Seront publiés en outre les articles traitant de l'anthropologie et de la morphologie clinique, en vue d'encourager la collaboration entre l'anatomie et les disciplines voisines.

Seront publiés en priorité les articles expressément demandés, nous tiendrons toutefois compte des articles qui nous seront envoyés dans la mesure où ils traitent d'un sejet dans son ensemble et correspondent aux standards des «Revues». Les publications seront faites en langues anglaise, allemande et française.

Dans l'intérêt d'une publication rapide et d'une large diffusion les travaux publiés paraitront dans des cahiers individuels, diffusés séparément: 6 cahiers forment un volume.

En principe, seuls les manuscrits qui n'ont encore été publiés ni dans le pays d'origine ni à l'éntranger peuvent nous être soumis. L'auteur s'engage en outre à ne pas les publier ailleurs ultérieurement. Les auteurs recevront 25 exemplaires gratuits de leur publication.

Manuscripts should be addressed to/Manuskripte sind zu senden an/Envoyer les manuscrits à:

Prof. Dr. A. BRODAL, Universitetet i Oslo, Anatomisk Institutt, Karl Johans Gate 47 (Domus Media), Oslo 1/Norwegen

Prof. W. HILD, Department of Anatomy, Medical Branch, The University of Texas, Galveston, Texas 77550/USA

Prof. Dr. J. van LIMBORGH, Universiteit van Amsterdam, Anatomisch-Embryologisch Laboratorium, Mauritskade 61, Amsterdam-O/Holland

Prof. Dr. R. ORTMANN, Anatomisches Institut der Universität, Lindenburg, D-5000 Köln-Lindenthal

Prof. Dr. T. H. SCHIEBLER, Anatomisches Institut der Universität, Koellikerstraße 6, D-8700 Würzburg

Prof. Dr. G. TÖNDURY, Direktion der Anatomie, Gloriastraße 19, CH-8006 Zürich/Schweiz

Prof. Dr. E. WOLFF, Collège de France, Laboratoire d'Embryologie Expérimentale, 49 Avenue de la belle Gabrielle, Nogent-sur-Marne 94/Frankreich

Advances in Anatomy, Embryology and Cell Biology
Ergebnisse der Anatomie und Entwicklungsgeschichte
Revues d'anatomie et de morphologie expérimentale

53/3

Dr. Faruk Hadžiselimović, Basler Kinderspital, Römergasse 8, CH–4000 Basle, Switzerland

To my Parents

ISBN-13: 978-3-540-08361-0   e-ISBN-13: 978-3-642-66715-2
DOI: 10.1007/ 978-3-642-66715-2

Library of Congress Cataloging in Publication Data. Hadžiselimović, Faruk, 1944-   . Cryptorchidism. (Advances in anatomy, embryology, and cell biology; v. 53/3) Bibliography: p.   Includes index. 1. Cryptorchism-Etiology. 2. Testicle-Growth. I. Title. II. Series. QL801.E67 vol. 53/3 [RJ477.5.C74] 574.4'08s [618.9'26'8] 77-24533

Composition: H. Stürtz AG, Universitätsdruckerei, Würzburg
2121/3321-543210

Contents

Preface

Cytological techniques have greatly improved in the last twenty years, largely as a result of further development of the microscope. Electron microscopy, in particular, has opened up great prospects for the study of cell morphology, while the development of radio-immuno-assay has brought great progress in endocrinology. The application of these two techniques, which are complementary to each other and provide mutually supporting evidence, is the subject of this monograph.

The work is divided into two parts. The first deals with the ultrastructure of normal testicular development, describing four main elements of the testicle — the germ cells, the Sertoli cells, the peritubular connective tissue and the Leydig cells — with details of their individual development. To make the electron micrographs more easily understandable, diagrams have been used to explain the most important points.

The second part deals with cryptorchid testicles, including the primary and secondary changes involved and morphometric studies of the secondary changes. The significance of these ultrastructural observations for the treatment of cryptorchidism is emphasized. With the aid of radio-immuno-assay the level of testosterone in cryptorchid mice was determined and comparisons drawn between the ultrastructural changes in cryptorchidism in mouse and man. The experimental studies served as a basis for the hypothesis that, in all probability, a congenital disturbance of the hypothalamo-hypophyso-gonadal axis is responsible for cryptorchidism.

I hope that this monograph will contribute towards a better understanding of normal and cryptorchid testicular development and of the etiology of cryptorchidism. In particular, I hope that the most important conclusion to be drawn, namely that it is imperative to treat cryptorchidism at an early age, will become evident to the reader.

Basle, February 1977   Faruk Hadžiselimović

Introduction

Cryptorchidism is a condition in which one or both testicles lies outside the scrotum
(Scorrer, 1964; Hoesli, 1971). According to Knorr (1970), cryptorchidism is the most
common endocrine gland disturbance. 2.7 % of all mature newborns are cryptorchid,
while the percentage of cryptorchids among premature infants is as high as 31 %.
(Scorrer and Farrington, 1971). The undescended testicle is or becomes permanently
damaged as a result of its situation, so that 60–85 % of all unilateral cryptorchids are
subfertile or sterile (Hansen, 1945; Michelson, 1952; Bayle, 1957; Carver, 1958;
Seguy, 1961; Guillon and Seguy, 1964; Hellinga, 1964; Nicole and Spindler, 1964).
In the case of bilateral cryptorchids, 91–100 % are subfertile or sterile (Hansen, 1945;
Michelson, 1952; Raboch and Zachor, 1955; Mack, 1960; Scott, 1961; Doepfmer and
Nienaber, 1964; Nicole and Spindler, 1964).

Already in the 19th century attempts were made to treat the consequences of cryptorchidism, namely the high rate of infertility. In 1887, Kocher published a monograph on disorders of the male sexual organs, introducing the era of surgical treatment
of cryptorchidism. The results, in terms of fertility, of surgically treated cryptorchids
were rather modest.

With the introduction of hormonal treatment for cryptorchidism, a better fertility
ratio might justifiably be expected (Schapiro, 1929). His observations that descent
could be induced by administration of gonadotropin were also confirmed, particularly
in experiments with macaques, by Hamilton (1939); Hamilton and Leonard (1938).

The time of operation was determined by the stage of development of histological
research. Thus, fifteen years ago, puberty was generally considered as the best time for
orchiopexy, the deciding factor in fixing the time for the operation being the histological changes in the seminiferous tubule, in particular the number of spermatogonia and
the diameter of the tubule. In 1971, Hoesli and Hedinger noticed that as early as the
third year there was a considerable decrease in the number of spermatogonia in cryptorchid as compared to normal testicles. On the basis of these observations, Hoesli and
Hedinger advocated early treatment of cryptorchidism, namely in the second year.
Since then, several researchers have been able to confirm the results of Hedinger and
Hoesli (Hecker, 1971; Städtler and Hartmann, 1972; Bodensky and Regele, 1973;
Jendricke et al., 1973).

The electron microscopic studies published before 1972 are not in agreement with
the light microscopic results. According to Leeson (1966), no morphological changes
are visible in the cryptorchid testicle until the tenth year. Numanoglu et al. (1969),
with the electron microscope, observed the first changes in Leydig cells and spermatogonia only in the fifth year.

The true causes of descensus and the reasons for a disturbance of this function have
remained a mystery until the present day. Raynaud (1942) made a decisive advance in
explaining the etiology of cryptorchidism, when he succeeded in producing the condition in the male issue of a mouse which had been administered estrogen during gestation. This experiment was repeated by Jean (1973), who came to the conclusion that
estrogen blocks androgen production by exerting a direct influence on the Leydig
cells, thus producing cryptorchidism. These experiments indicate an endocrinological
factor in the etiology of cryptorchidism. In this connection, it is important to point
out that Hayashi and Harrison (1971), with the light microscope, were unable to find

any Leydig cells in the interstitium of one-year-old cryptorchids. They also observed retarded development of the Leydig cells of cryptorchids in puberty. These findings were also confirmed on an endocrinological basis by Raboch and Zachor (1955), who discovered reduced androgen secretion in cryptorchid Leydig cells in young adults. A reduced testosterone content in cryptorchid testicles was also observed in the pig (Hanes and Hooker, 1937) and in the dog (Eik-Nes, 1966). From here it is thus a short step to the theory that an endocrinological disturbance during intrauterine development lies at the bottom of cryptorchidism.

Another group of researchers ascribes the occurrence of cryptorchidism to a congenital disturbance of the spermatogonia (Farrington and Scorrer, 1971). According to these authors, cryptorchidism and the high infertility rate are the result of "low quality" spermatogonia.

The third possible cause of cryptorchidism could be a malformation in the inguinal passage. This theory that the etiology of cryptorchidism is of a mechanical nature has been repeatedly put forward, particularly by surgeons and pathologists.

Material and Methods

1. Biopsies from Child Testicles

Ultrastructural studies were carried out on 154 biopsies obtained from the cryptorchid testicles of children between the ages of two weeks and sixteen years (Table 1). 37 biopsies of normal testicles from children aged between a few days and fourteen years served as a control. Most of the biopsies from normal testicles, particularly up to the age of one, were obtained immediately post mortem.

Table 1. Biopsy material

Age	Cryptorchid testis	Normal testis
1	13	12
2	16	6
3	13	3
4	6	1
5	14	1
6	11	2
7	12	2
8	13	1
9	11	1
10	15	1
11	13	1
12	5	1
13	6	2
14	2	2
15	4	1
total:	154	37

These children had either died suddenly in infancy or from illnesses which are known to have no effect on the testicles, such as osteogenesis imperfecta, subdural or intraventricular hematoma. The older children, i. e. those between three and eight, were either biopsied immediately after death (mostly the result of accidents) or during an operation for inguinal hernia. The puberty age-group can be divided into two sub-groups, the first consisting of patients where biopsies were taken from both testicles, because of premature enlargement of one testicle. These were cases of suspected malignoma, which routine histological examination later revealed to be normal. The smaller group – three in all – comprised biopsies obtained in the course of operations for inguinal hernia.

2. Animal Experiments

For ultrastructural observations, the newborn mice were divided into three groups, each of twelve mice. One group served as control, the second group was made up of newborn mice whose mother had received 5 mg of estrogen (Ovocyclin (Ciba)- Oestradiolmonobenzoat) on the 14th day of gestation, while the third group comprised mice whose mother had received 30 I.U. HCG in addition to 5 mg of estrogen on the same day of gestation. The testosterone content of whole testicles was determined by the direct radio-immuno-assay method on fourteen untreated newborns and 19 newborns whose mother had received 5 mg of estradiol.

The adult mice were divided into two groups. One group of eleven mice comprised those who had been treated i. u. with estradiol. The second group of six mice served as control. Before homogenising the testes, in order to determine the testosterone content by radio-immuno-assay, biopsies were taken for electron microscopic studies.

The homogenates from newborn and adult mice testes were extracted in 1 ml ether. A direct radio-immuno-assay was performed according to Ismail et al using an I^{125} labeled testosterone. (The specific antiserum was kindly donated by A. A. A. Ismail, Endoc. Res. Unit, Edinburgh.) The intraassay coefficient of variation was 5 %. Extracts from controls in $E_2 B$ mice were estimated in the same assay.

The biopsies were fixed in 3 % glutaraldehyde containing PBS, postfixed in 1 % OsO_4 and dehydrated in alcohol and propylenoxide. The tissue blocks were then embedded in Epon 812. The silver sections were cut with an LKB ultramicrotome, stained with uranyl acetate and lead citrate and examined with a Zeiss EM 9A electron microscope.

The Normal Testicle

1. General Observations on the Development of the Normal Testicle in Children

The development of the testicle in children is a continuous process, which goes on after birth until puberty. The diameter of the seminiferous tubule increases steadily until the 12th year (Fig. 14). Only with the appearance of the lumen does a sudden rapid increase in diameter take place.

The testicle is composed mainly of four elements: germ cells, Sertoli cells, peritubular connective tissue and Leydig cells. No mention will be made in this monograph of the innervation nor of the lymph and blood vessels of the testicle.

a) The First Year of Life (Fig. 1)

Immediately after birth, the seminiferous tubule is composed of gonocytes, spermatogonia and Sertoli cells. The gonocytes or primitive reproductive cells are located

Table 2

Morphometrical symbol	Normal testis		Cryptorchid testis	
	$\bar{\bar{x}}$	S.E.	$\bar{x}$	S.E.
Number of Sertoli cell nuclei cu/cm	1461×10^6	134×10^6	1617×10^6	103×10^6
Single cell volume cu/μ	491	55	429	17
Volume density of spermatogonia cu/cm	0 056	0 018	0 021	0 009
Volume density of degenerating cells	0 012	0 003	0 018	0 006

mainly in the centre of the tubule and show a tendency to move towards the basement membrane. Two different types of gonocytes can be distinguished. When the gonocyte comes into contact with the basement membrane, it changes into a spermatogonium, the largest cell in the infant seminiferous tubule. The most common cell in the seminiferous tubule in the first year is the Sertoli cell, an oval or polarised cell which, according to definition, is always in contact with the basement membrane. It is a small cell, (491 cu/μ) fulfilling, in addition to phagocytizing, hormone-producing and nutritive functions, a supporting role.

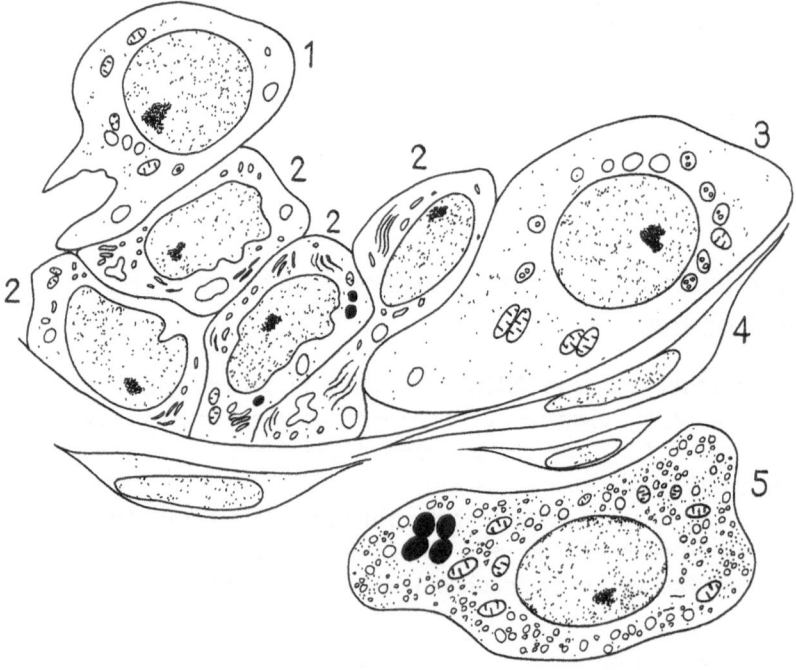

Fig. 1. Diagram of a 2-month-old normal testicle. *1* Gonocyte, *2* Fetal Sertoli cells (S_f), *3* Fetal spermatogonia, *4* Fibroblasts, *5* Leydig cells

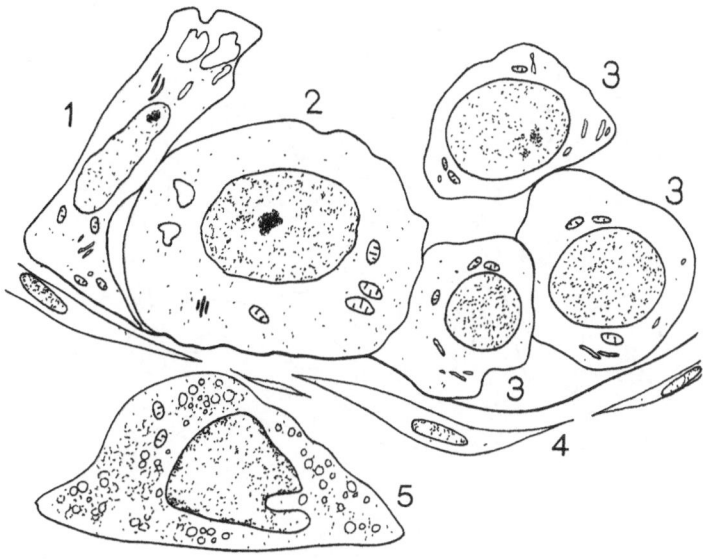

Fig. 2. Diagram of a 4-year-old normal testicle. *1* S$_b$ Sertoli cells, *2* Fetal spermatogonia, *3* S$_a$ Sertoli cells, *4* Fibroblasts, *5* Leydig cells

The peritubular connective tissue, which forms the wall of the seminiferous tubule, is composed of the basement membrane, here consisting of one layer, a collagen fibre zone and fibroblasts. The fibroblasts form concentric rings around the tubule. The interstitium contains mainly fetal Leydig cells, which are well developed and can be found in the interstitium until the second year.

b) The Fourth Year (Fig. 2)

The seminiferous tubule has an ultrastructural appearance quite different from that of a one-year-old. Gonocytes are no longer visible. In addition to the A-type spermatogonia, which can already be seen in the one-year-old, B-type spermatogonia are encountered for the first time. Simultaneously with the appearance of B-type spermatogonia, primary spermatocytes are found in the seminiferous tubule. The Sertoli cells have completed their transformation from fetal cells into Sa- and Sb-type cells. The Sa-type cell is the most common of the Sertoli cells in the seminiferous tubule in children. In the fourth year, simultaneously with the appearance of the B spermatogonia and primary spermatocytes, Sb-type Sertoli cells are found in increasing numbers.

Apart from a certain widening, there are no qualitative changes in the peritubular connective tissue. The basement membrane still consists of one layer and no knob formation is discernible. The collagen fibre layer is wider and the cellular layer is composed of fibroblasts. The interstitium contains mainly precursors of Leydig cells; occasionally, however, particularly between the ages of four and eight, juvenile Leydig cells can be seen, grouped around the vessels.

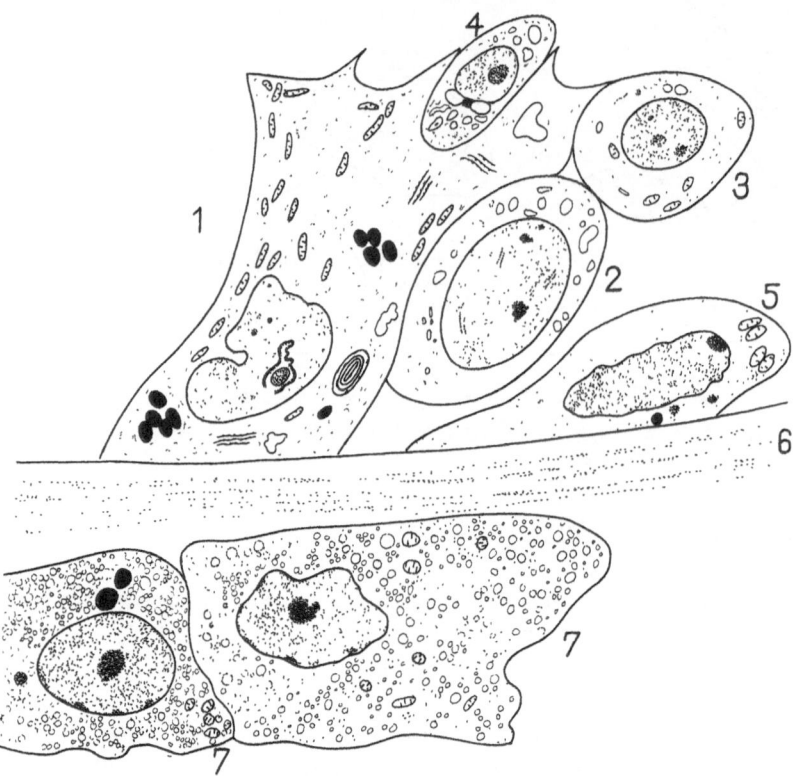

Fig. 3. Diagram of a 13-year-old normal testicle. *1* S$_c$ Sertoli cells, *2* Primary spermatocytes, *3* Secondary spermatocytes, *4* Spermatids, *5* A$_p$ spermatogonia, *6* peritubular connective tissue, *7* Leydig cells

c) Puberty (Fig. 3)

In puberty the final development of all four elements is completed. The Sertoli cells increase to as much as five times their previous size and the transition to the Sc-type cell takes place. The number of Sertoli cells has been steadily decreasing from birth to puberty. The seminiferous tubule acquires a lumen and degenerating cells are very seldom seen. The germ cells comprise spermatogonia, primary spermatocytes, secondary spermatocytes and spermatids. Up until the age of fourteen, it was impossible to locate any sperm in our material. Only when the Sertoli cells are fully developed, do the sperm appear. In the peritubular connective tissue a change takes place under the influence of gonadotropin. The basement membrane becomes multilayered and a knob formation becomes visible. The collagen fibres run in an orderly fashion and a transformation from fibroblasts to myofibroblasts takes place. The seminiferous tubule thus acquires a certain contractility.

The Leydig cells are very well developed, with a particularly noticeable increase in the amount of smooth endoplasmic reticulum. However, no Reinecke's crystalloids are as yet visible.

2. Germ Cells

Two types of germ cells can be distinguished in the seminiferous tubule of the newborn: gonocytes and spermatogonia.

a) Gonocytes (Fig. 4)

The gonocyte or primordial germ cell according to by Sapsford (1962) is defined as a cell from which the spermatogonia originate by transformation and having the capacity to move in an ameboid manner. According to McKay et al. (1953), gonocytes give a positive reaction to alkaline phosphatase.

The ultrastructure of fetal gonocytes has been described in detail by several authors (Gondos and Hobel, 1971; Wartenberg et al., 1971; Fukuda et al., 1975).

The gonocyte of the newborn corresponds to the second phase gonocyte described by Wartenberg el al., 1971 and the fetal gonocytes of Gondos and Hobel (1971) and Fukuda et al. (1975). These latter describe a second type of gonocyte in the fetal seminiferous tubule, a transitional stage between gonocyte and spermatogonium, which they call an "intermediate cell". The intermediate cells occurring in the fetal testis are reported to be somewhat smaller than the gonocytes and to possess crista-type mitochondria (Fukuda et al., 1975). This type of cell can also be observed in the seminiferous tubule of newborns.

The gonocytes are smaller than the spermatogonia and are localised mainly in the centre of the tubule. In the last months i. u. and the first weeks post natum, they migrate to the periphery (Gondos and Hobel, 1971). As soon as they reach the basement membrane, they give rise to fetal spermatogonia. When the gonocyte begins to move towards the basement membrane, long pseudopodia-like processes extend towards the periphery of the tubule. The round nucleus has one to two nucleoli, which are centrally situated and have a loose reticulum.

The chromatin forms small clumps and is less dispersed than that of the fetal spermatogonia. Occasionally, a protrusion can be observed on the nuclear membrane. This cap-like formation is a marked lifting of the nuclear membrane from the structure, with a cavity in the interior and is evidently a normal phenomenon (Fukuda et al., 1975; Hadžiselimović, 1976). Such nuclear protuberances can also be observed in gonocytes where no signs of any degeneration are visible. This structure on the nucleus is connected with the appearance of the so-called chromatid bodies, which are frequently present in the cytoplasm of fetal spermatogonia (Fukuda et al., 1975; Hadžiselimović, 1976).

The narrow cytoplasm contains a relatively large number of small, round mitochondria of the tubule type. These are generally larger than those of the Sertoli cells and, as a rule, do not form any intermitochondrial substance. A large number of polyribosomes and glycogen granules are visible in the cytoplasm. The rough endoplasmic reticulum and the smooth endoplasmic reticulum are very sparse. The pseudopodia-like gonocytic processes, which extend towards the periphery, contain micropinocytotic vesicles.

A second type of gonocyte, which is generally smaller and whose nuclear cytoplasm is displaced in favour of the nucleus, also exists in the neonatal seminiferous tubule. The cytoplasm is electron-light, with a small amount of rough endoplasmic reticulum and crista-type mitochondria. Glycogen is completely lacking in the cytoplasm of this

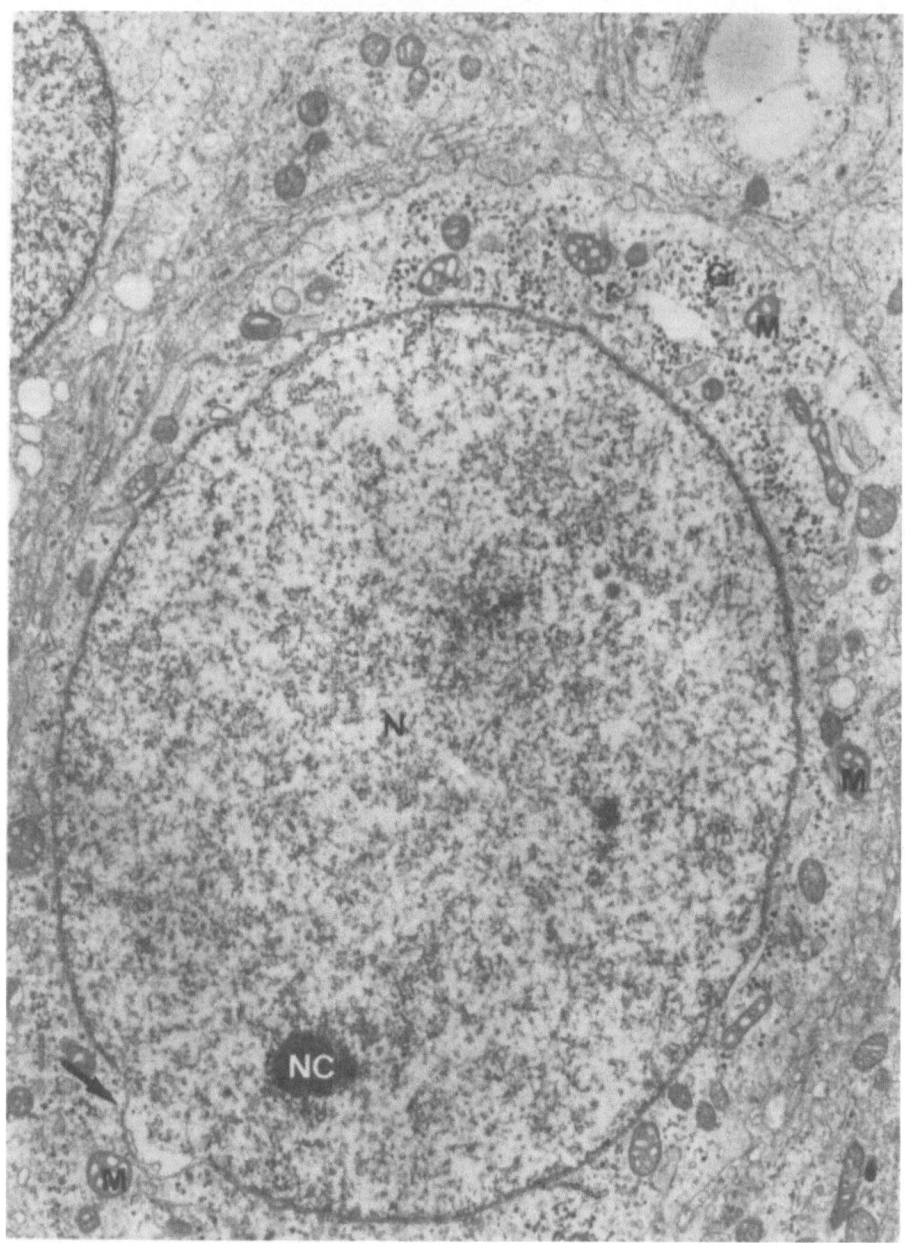

Fig. 4. Gonocyte of a 5-week-old boy. *G* glycogen, *M* mitochondria, *N* nucleus, *NC* nucleolus, (↑) indicates cap formation. X 30,000

gonocyte. The long pseudopodia, which, in the first type of gonocyte described above, are directed towards the basement membrane, are here only rudimentary, or completely absent.

From the ultrastructural point of view, these cells may be said to correspond to the so-called "intermediate cells" (Fukuda et al., 1975). The fact that these intermediate

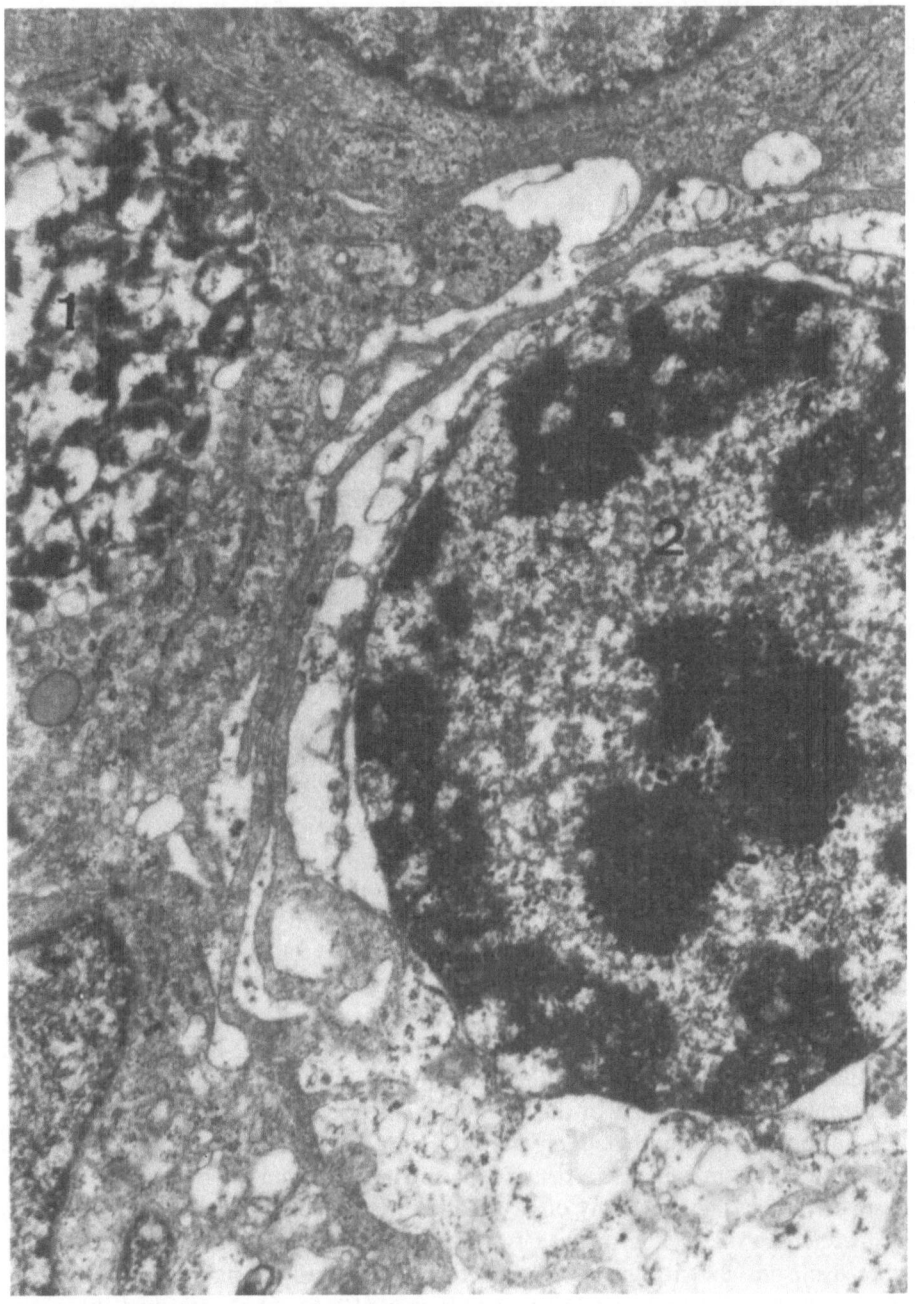

Fig. 5. Gonocyte degeneration in the seminiferous tubule of a 5-week-old boy. Two distinct stages of degeneration can be differentiated *2* Early – the entire nucleus is still visible, and *1* Late – only remnants of a gonocyte can be seen. X 25,000

cells are found mainly in the centre of the tubule, possess no or only very rudimentary pseudopodia, a narrow, light cytoplasm and exhibit signs of degenerating mitochondria, supports the assumption that they are gonocytes which will later be phagocytized by the Sertoli cells (Hadžiselimović, 1976). In the centre of the tubule, degenerating gonocytes are frequently found. The Sertoli cell plays an important role in the disposal of the degenerating gonocytes, surrounding them with its pseudopodia and drawing them into its cytoplasm (Fig. 5). The phagocytized gonocytes are digested in the cytoplasm of the Sertoli cells and appear as light vacuoles.

The gonocytes which advance as far as the basement membrane become transformed into fetal spermatogonia. The entire gonocyte becomes enlarged, the cytoplasm showing the greatest development. The mitochondria form connections with each other by means of the intermitochondrial substance. Fetal spermatogonia have never been observed to arise as a result of cellular division of the gonocytes, which can be seen in the seminiferous tubule of newborns until the third month post natum.

b) Spermatogonia

Spermatogonia are described as cells which are in contact with the basement membrane and from which primary spermatocytes develop by cell division. They were first observed by La Valette St. George (1876). Until recently, it was generally believed that the final spermatogonia did not develop from the gonocytes until puberty (Monesi, 1972). Vilar (1970) distinguished two types of germ cells in children, on an ultrastructural basis: firstly the gonocytes and secondly the primitive spermatogonia. These are reported to decrease in number until puberty, at which time they develop into adult spermatogonia (Vilar, 1970).

My own observations have shown that from birth to puberty, in addition to gonocytes, fetal spermatogonia, transitional spermatogonia, Ap, Ad and B spermatogonia, as well as primary spermatocytes, can be discerned (Fig. 6).

Fetal Spermatogonia (Fig. 7): Fetal spermatogonia develop from the gonocytes and are situated, according to reports, on the basement membrane. They have a large cell body with a noticeably electron-light cytoplasm. The form of the nucleus is round to oval, about twice as large as that of the neighbouring Sertoli cells. In the middle of the nucleus lies a large nucleolus, consisting of loose reticular substance and two to three amorphous bodies. The chromatin is distributed over the entire nucleus in the form of fine granules. A typical feature of the spermatogonia is the large mitochondria, which are connected with each other by the intermitochondrial substance. They are mainly of the tubule type, but a few of the crista type can also be observed. The Golgi complex and rough and smooth endoplasmic reticulum are sparsely distributed in the electron-light cytoplasm, which is arranged concentrically around the nucleus. So-called chromatid bodies are frequently observed in the vicinity of the nucleus.

The fetal spermatogonia are the largest cells in the child seminiferous tubule and can be distinguished with the electron microscope up until the sixth year. This corresponds with the observations of Niemi and Ikonen (1965), who, with the light microscope, were able to distinguish the fetal spermatogonia clearly from the other spermatogonia, because of their different histological activity (positive alkaline phosphatase reaction) and who were able to establish their presence up to the seventh year.

Transitional Spermatogonia (Fig. 8): Transitional spermatogonia originate from fetal spermatogonia in the first six years of life (Seguchi and Hadžiselimović, 1974). They are, on the whole, somewhat smaller than the fetal spermatogonia and are described

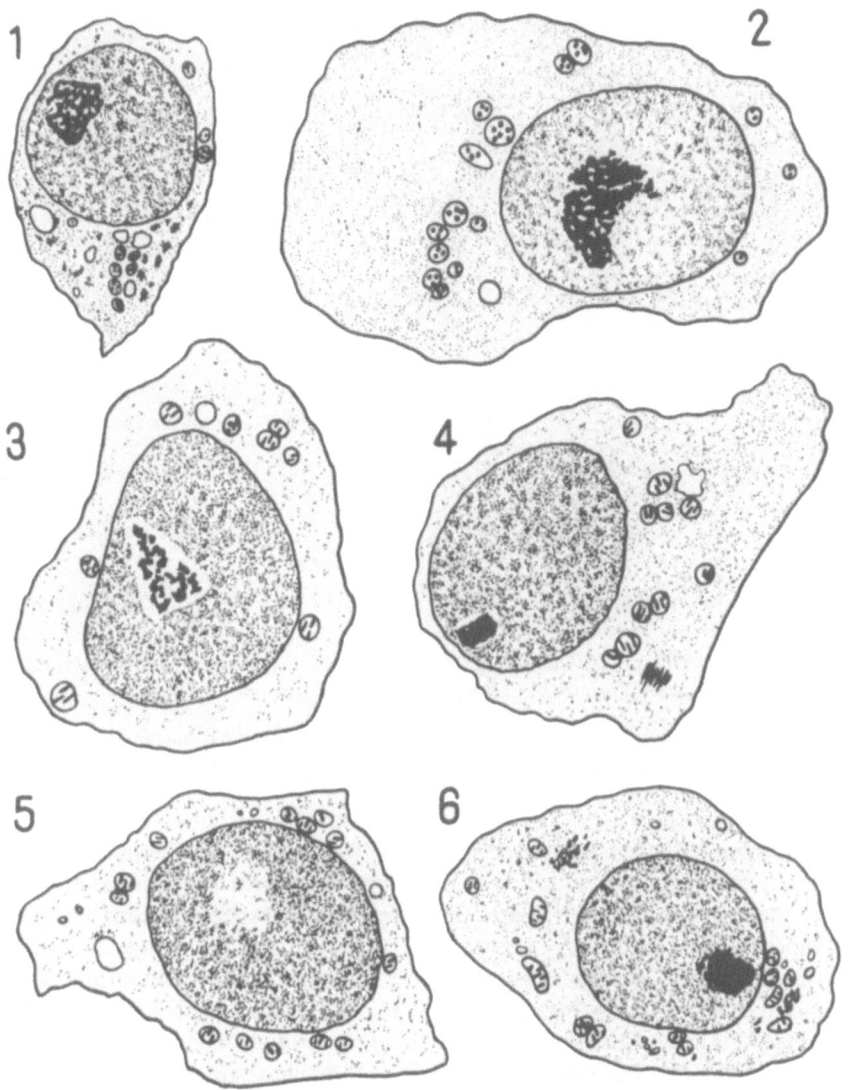

Fig. 6. Various types of germ cells. *1* Gonocyte, *2* Fetal spermatogonia, *3* Transitional spermatogonia, *4* A_p spermatogonia, *5* A_d spermatogonia, *6* B spermatogonia

as being attached to the basement membrane. Their nucleus is oval and the nucleolus, which is no longer centrally, but rather peripherally situated, consists of loose reticular material and is surrounded by a light halo. The chromatin is distributed finely over the entire nucleus. The mitochondria are connected with the nuclear membrane and are mainly of the the crista type, joined together by intermitochondrial substance. The cytoplasm of the transitional spermatogonia appears to be electron denser than that of the fetal spermatogonia and its rough and smooth endoplasmic reticulum is generally sparsely distributed. Glycogen granules are more frequently found than in the fetal spermatogonia.

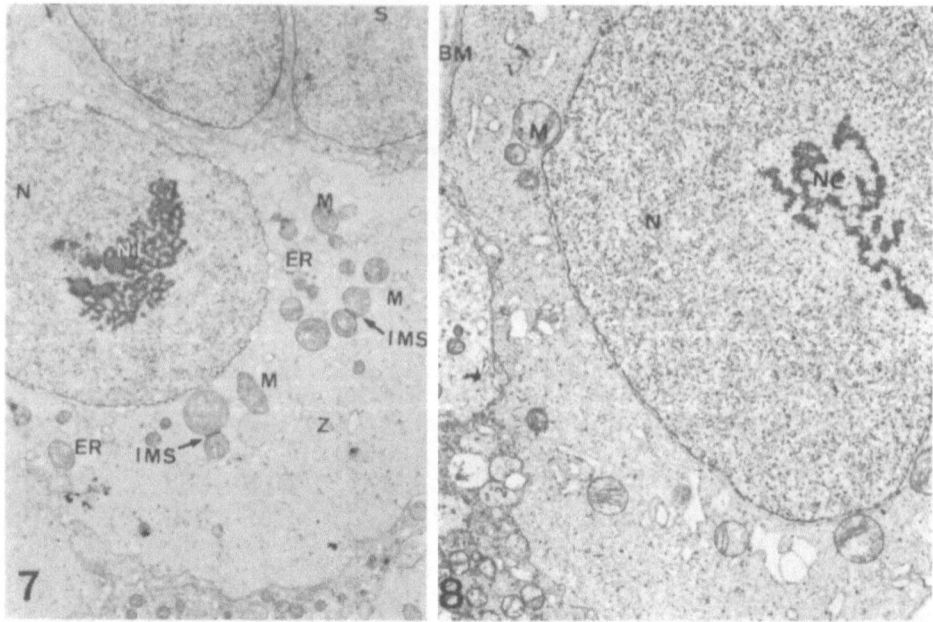

Fig. 7. Fetal spermatogonium. *ER* endoplasmic reticulum, *IMS* intermitochondrial substance, *M* mitochondria, *N* nucleus, *NL* nucleolus, *S* S$_a$ Sertoli cells, Z Cytoplasm X 4,900

Fig. 8. Transitional spermatogonium from a 2-year-old normal testicle. *NC* nucleolus moving towards the periphery of the nucleus. *BM* basement membrane, *M* mitochondria. X 6,300

Ap Spermatogonia (Fig. 9): As early as five weeks post natum, Ap spermatogonia are present in the infant seminiferous tubule. This type of spermatogonia is the most common until the age of fourteen. It develops from transitional spermatogonia and is two to four times as big as the Sa-type Sertoli cell. The largest Ap spermatogonia are approximately 30 μ long. The nucleus is excentrically situated and round or polygonal in form. So-called "bleb" formations (Rowley et al., 1971) can also be observed on the nucleus (Fig. 9). The chromatin is fine and homogeneously distributed. One to three nucleoli, consisting of a fine granular, homogeneous nucleus, surrounded by loose reticular substance, can be seen attached to the nuclear membrane.

No mitochondria are observed in contact with the nuclear membrane. The cytoplasm is of the fine granular type and generally has a few uncharacteristic cell organelles. Occasionally, large vacuoles can be seen in the cytoplasm. They are empty and bounded by a membrane. The mitochondria are of the crista type, grouped around the nucleus and connected to each other by intermitochondrial substance. The endoplasmic reticulum is in contact with the mitochondria. The slightly developed Golgi apparatus is situated in the vicinity of the nucleus. Ribosomes, microtubules and glycogen granules are distributed throughout the cytoplasm.

The crystalloid of Lubarsch (1896) can be demonstrated only in A and B type spermatogonia. They are always found in the vicinity of the nucleus, frequently surrounded by clusters of mitochondria and endoplasmic reticulum. In the testicles of thirteen-year-olds, the crystalloid is approximately 5.9 μ long and approximately 0.5 μ wide. It consists of parallel fibrils, ribosome-like granules and fine granules. The fibrils

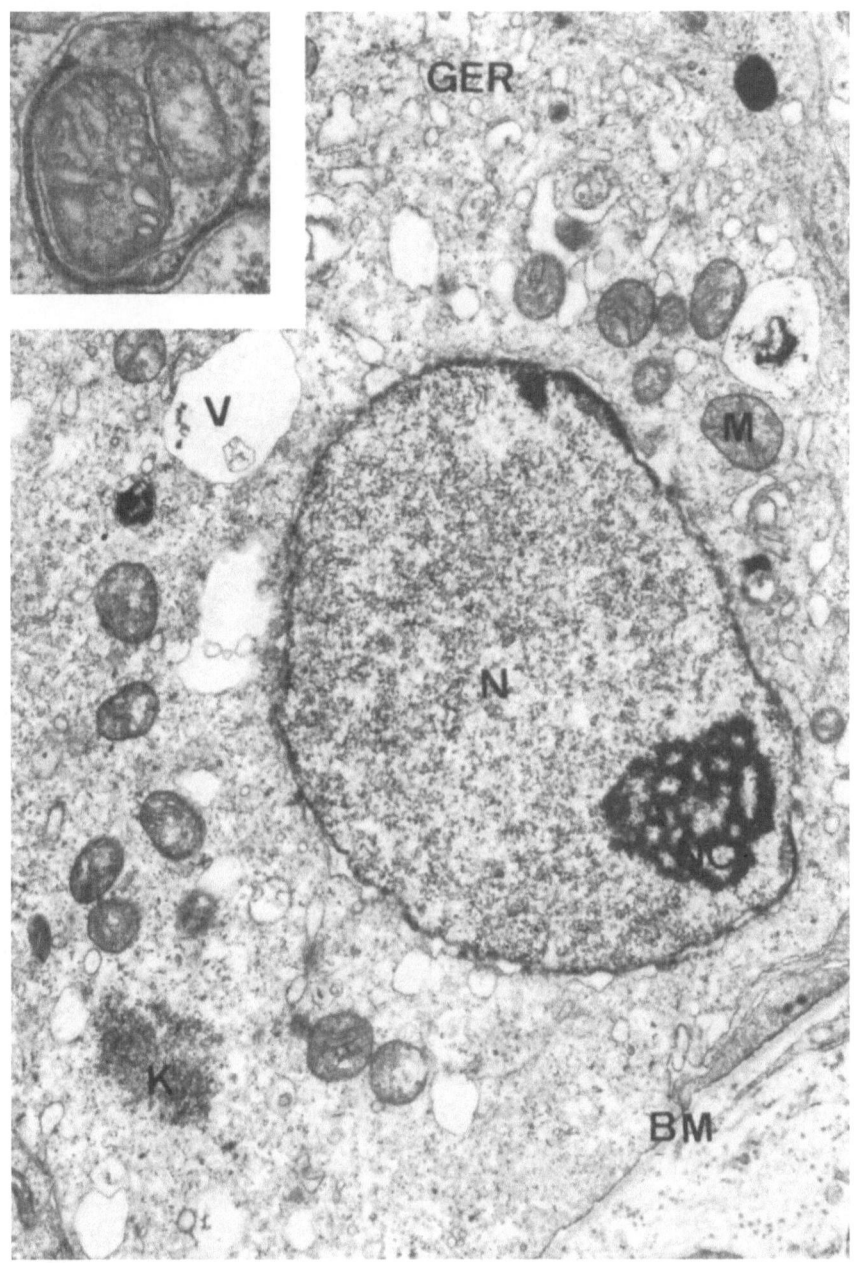

Fig. 9. A$_p$ spermatogonium of a 5-week-old boy. *BM* basement membrane. *GER* smooth endoplasmic reticulum, *K* chromatid body, *M* mitochondria, *NC* nucleolus, *N* nucleus, *V* vacuoles. X 14,000. Enlarged section: "Bleb" formation of an Ap spermatogonium. In the interior two mitochondria can be seen — this indicates a nuclear invagination in this region

are approximately 80–150 Å thick, running parallel to each other, with a space of about 100–300 Å between. They stretch almost without interruption from one end to the other. The ribosome-like granules, with a diameter of approximately 100–200 Å,

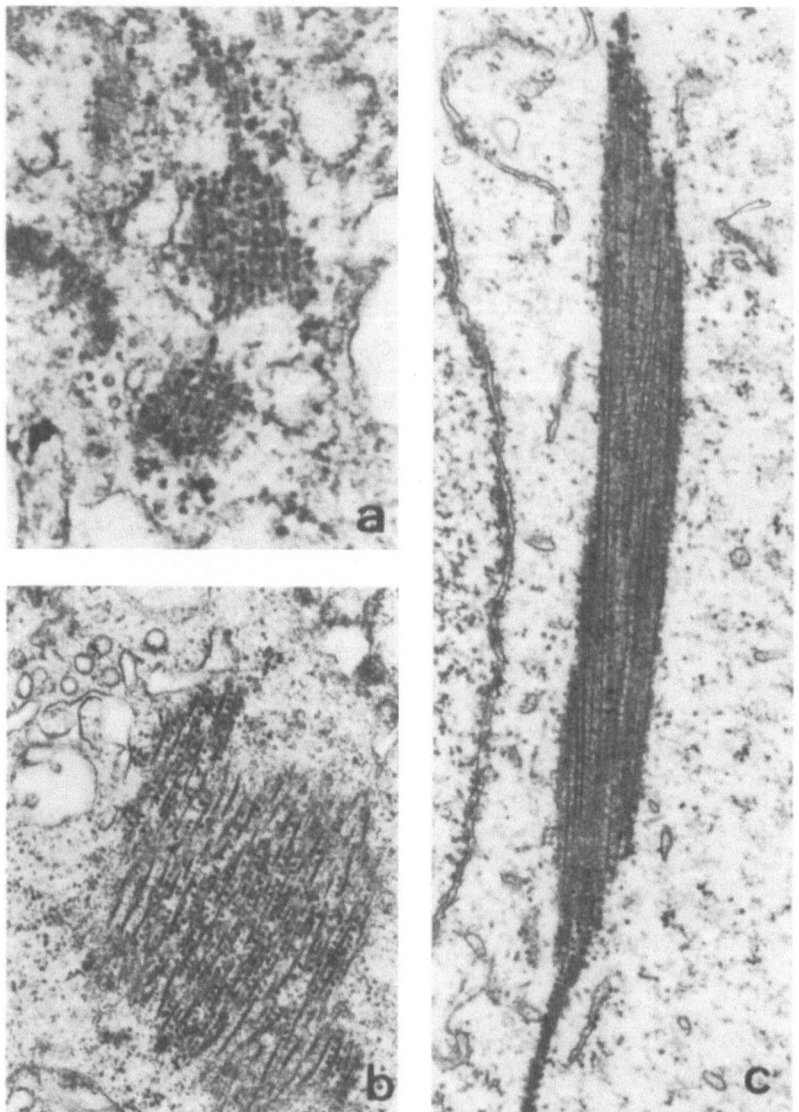

Fig. 10 a–c. Crystalloid development. (a) 4-month-old testicle. X 90,000. (b) 6-year-old testicle. X 75,000. (c) crystalloid in the spermatogonium of a 13-year-old boy. X 25,000

are found between the fibrils. They are generally located peripherally in the testicles of thirteen-year-olds.

Ribosome-like granules are also frequently found in the immediate surroundings of the crystalloid, which has a basically fine granular structure, with occasional lighter zones and no limiting membrane. Isolated crystalloid particles are also visible in the vicinity of the nucleus. These are probably bent or branching crystalloids. The appearance seems to depend on the direction of the section. The largest of these particles is approximately 1 μ long and 0.6 μ wide.

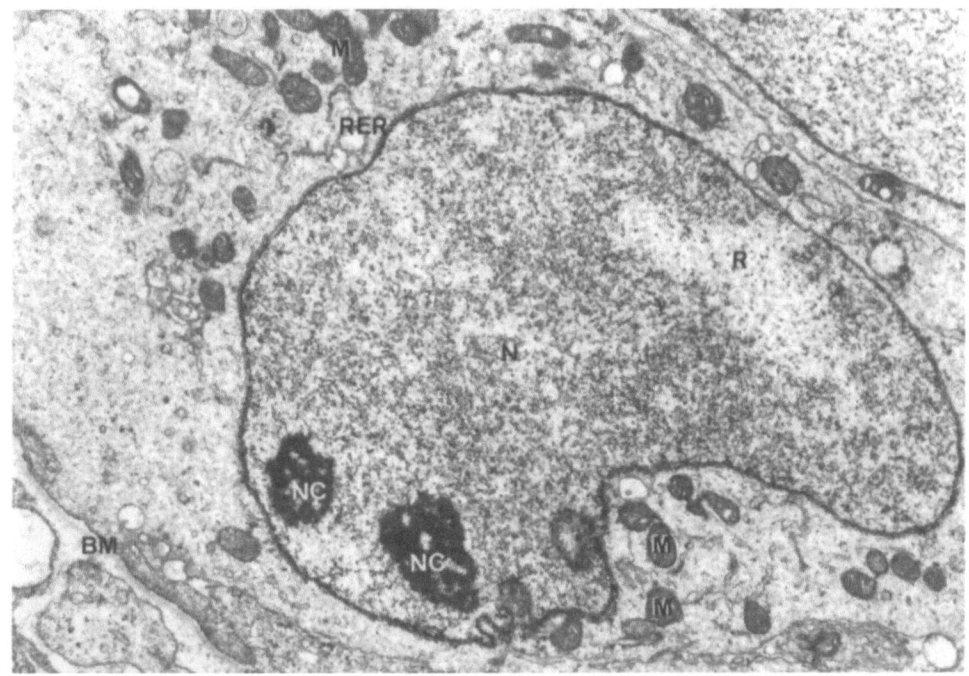

Fig. 11. A$_d$ spermatogonium of a 5-week-old normal testicle. The nucleus N has a typical lighter zone R. The nucleolus NC lies on the periphery. M The mitochondria are sometimes in contact with the nuclear membrane. RER rough endoplasmic reticulum. BM basement membrane. X 13,500

Unlike the previously described ultrastructure, the crystalloid in one-year-olds consists of short fibrils. In the intermediate spaces between the individual fibrils, throughout the entire crystalloid, are found large granules, lying parallel to each other. The basic structure of the crystalloid also exhibits fine granules, which are visible around the crystalloid too.

In isolated A spermatogonia and in fetal spermatogonia, peculiar structures can be observed in the vicinity of the nucleus. Some of these are fine granules, surrounded by ribosome-like granules. Although no fibrillary structures are visible, one has the impression that this structure is beginning to assume a fibrillary arrangement. This is the so-called "chromatid body". Another formation is also visible, with rather irregularly running fibrils and fine granules. The ribosome-like granules between the fibrils are, however, absent.

Ad Spermatogonia (Fig. 11): Ad spermatogonia are the second commonest type of spermatogonia in the infant seminiferous tubule. Their characteristic feature is a lighter zone, the so-called "rerarification". This zone is generally located in the middle of the nucleus. The nucleoli are arranged peripherally, but have rarely if ever any contact with the nuclear membrane. The crista-type mitochondria are connected with the nuclear membrane and often with each other by intermitochondrial cement. The Ad spermatogonia generally have more rough endoplasmic reticulum than the Ap-type. This is found in the cytoplasm, which is arranged concentrically around the nucleus. As in the Ap-type, here too, Lubarsch's crystalloid is situated in the cytoplasm, in which isolated ribosomes, microtubuli and glycogen granules are also to be found. The young Ad spermatogonia can be observed as early as five weeks post natum and often have nuclear invaginations. In thirteen-year-olds, the nucleus of the Ad spermatogonia is oval and has no such invaginations.

B Spermatogonia (Fig. 12): B spermatogonia are smaller and more rounded than the A-type, from which they develop. They have a very small area of contact with the basement membrane. The round nucleus is situated excentrically in the cytoplasm. The chromatin is no longer homogeneously distributed, but forms granules. The nucleolus lies on the periphery of the nucleus and has no contact with the nuclear membrane, which is in contact with the mitochondria. The well developed Golgi complex lies in the vicinity of the nucleus. A lamellar and a vesicular part can be distinguished. The endoplasmic reticulum, both the rough and the smooth, is much better developed than that of the A spermatogonia. In the cytoplasm polyribosomes and, as in the A-type, Lubarsch's crystalloid, are also present. Crista-type mitochondria are found generally singly in the cytoplasm; very rarely they are connected to each other by intermitochondrial substance. The "bleb" structure observed in the nucleus of A-type spermatogonia was not found in the B-type.

Rowley et al. (1971) described the ultrastructure of spermatogonia in adults, dividing them into two groups: A and B spermatogonia. Type A spermatogonia were further subdivided into Ap (pale), Ad (dark) and Al (long).

Unlike Vilar (1970), we were able to establish the presence in children of Ap, Ad and B spermatogonia, as described by Rowley et al. (1971), in addition to fetal and transitional spermatogonia (Hadžiselimović and Seguchi, 1975; Seguchi and Hadžiselimović, 1974). The A and B spermatogonia in the child fulfil the conditions set down by Rowley et al. (1971) for these types of spermatogonia in adults.

Spermatogonia counts carried out on semi-thin sections of a representative number of 30 tubuli per biopsy showed on tubulus cross-sections that the number of spermatogonia increases until the age of twelve. A total of twelve biopsies were counted (Seguchi and Hadžiselimović, 1974). The increase in spermatogonia apparently proceeds in linear fashion until the twelfth year. Between the age of twelve and thirteen, not only the number of spermatogonia but also the diameter of the seminiferous tubule shows a rapid increase (Fig. 14). In the first four years of life, the average is one spermatogonium per tubule. In the five- to eight-year-old group, the average is 2.2 and in the nine- to twelve-year-olds, this increases to 2.5 spermatogonia per tubule. The difference between the first group (one- to four-year-olds) and the third group (nine- to twelve-year-olds) is significant ($P < 0.05$). In the thirteen-year-old group, there is a mean of 18 $\frac{1}{2}$ germ cells per cross-section of tubule (Table 2).

c) Primary Spermatocytes (Fig. 13)

Simultaneously with the appearance of B spermatogonia in the fifth year, primary spermatocytes can be distinguished. They were identified in four- and five-year-olds by Scorer and Farrington (1971) using the light microscope. In our material, primary spermatocytes were identified in four-, five-, six- and seven-year-olds, after which no more primary spermatocytes were found until the thirteenth year, when they began to appear in increasing numbers. Characteristic of primary spermatocytes is that they have no contact with the basement membrane and are surrounded by Sertoli cells. In

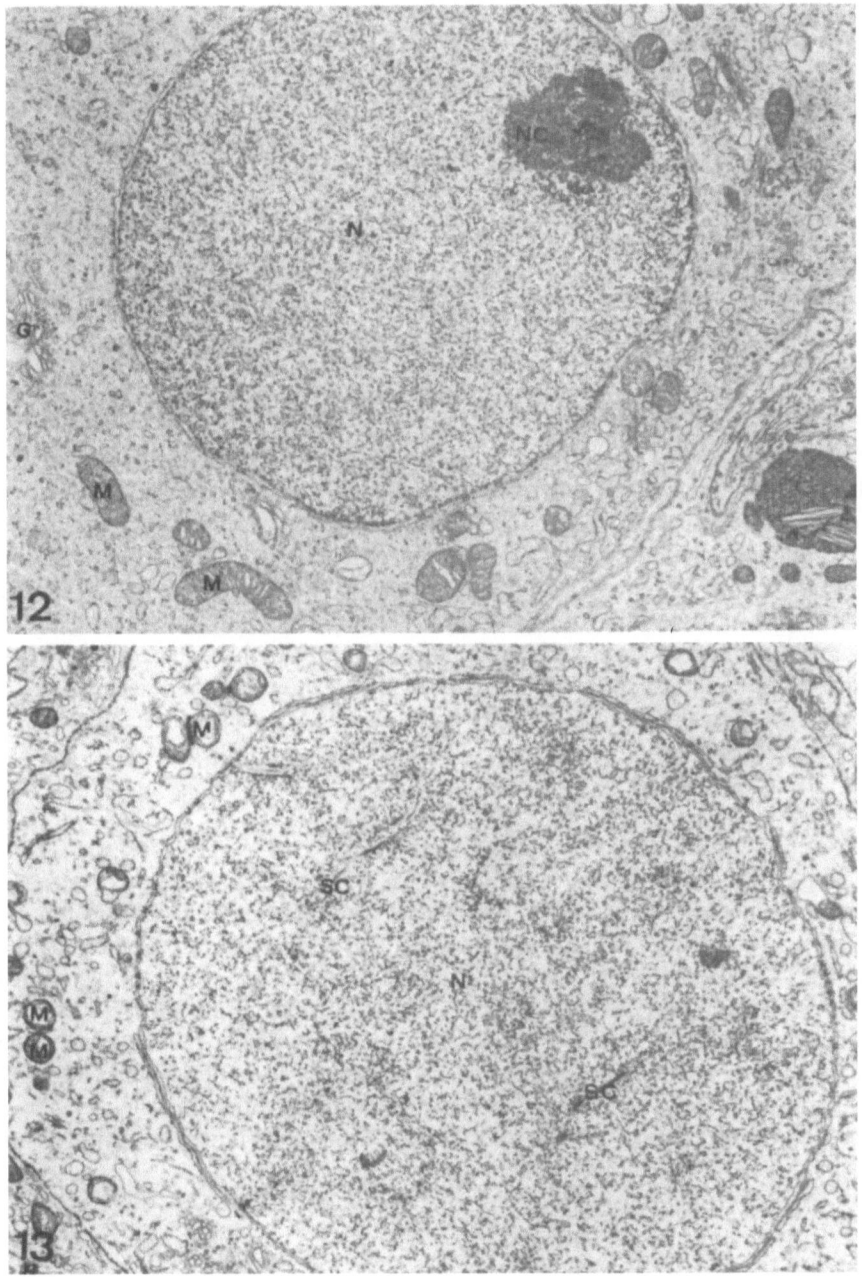

Fig. 12. B spermatogonium. *G* Golgi apparatus, *M* mitochondria, *N* nucleus, *NC* nucleolus. X 8,000

Fig. 13. Primary spermatocyte. *M* mitochondria, *N* nucleus, *SC* synaptilemal complex. X 7,200

thirteen-year-olds, intracellular bridges were particularly common in primary spermatocytes, which are approximately the same size as B spermatogonia. The ultrastructure reveals a round nucleus, centrally located in the electron-light cytoplasm. The nucleus has one to three nucleoli, which are rarely connected with the nuclear membrane. The

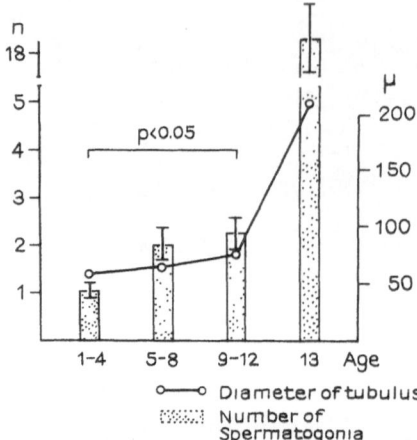

Fig. 14. Development of the diameter of the seminiferous tubule and the spermatogonia count

Diameter of tubulus
Number of Spermatogonia

nucleus has irregularly dispersed chromatin and granules and synaptilemal complexes typical of primary spermatocytes. The synaptilemal complexes correspond to those in adult primary spermatocytes. The nuclear membrane is in contact with the endoplasmic reticulum. The rough endoplasmic reticulum forms conglomerates in various parts of the cytoplasm. The mitochondria have a peculiar appearance, their cristae being raised and there being no arrangement of cristae or tubules. In size, they are similar to those of the Sertoli cells and much smaller than those of the spermatogonia. Occasionally, they are joined together by intermitochondrial substance. Spermatogenesis remains at the stage of primary spermatocytes until puberty.

d) Degenerated Cells (Fig. 15)

In the seminiferous tubule, which in children has no lumen, degenerating or degenerated cell zones are sometimes found. This is peculiar to the child seminiferous tubule. The degenerated cells are mainly spermatogonia and Sertoli cells. Most of the degenerating cells lie in the interior of the tubule, but occasionally such zones are also found on the periphery. Inside this zone, the cells have lost their limiting membrane and nuclear remnants can be seen everywhere. The isolated nuclei are markedly pyknotic on the periphery and are in the process of disintegration. The mitochondria lose their internal structure and a large number of isolated vacuoles, with a limiting membrane, can be seen throughout the zone. With the appearance of the lumen in the thirteenth year, the degenerating cells disappear.

3. Sertoli Cells

The Sertoli cells were named after their discoverer, Sertoli, who was the first to describe them in 1865. This cell is polymorphous in appearance, with a characteristic, irregular nucleus in adults. The Sertoli cell is always in contact with the basement membrane. In addition to its supporting function, nutritive, hormone-producing, transport-regulating and phagocytizing roles are ascribed to it. The literature contains nu-

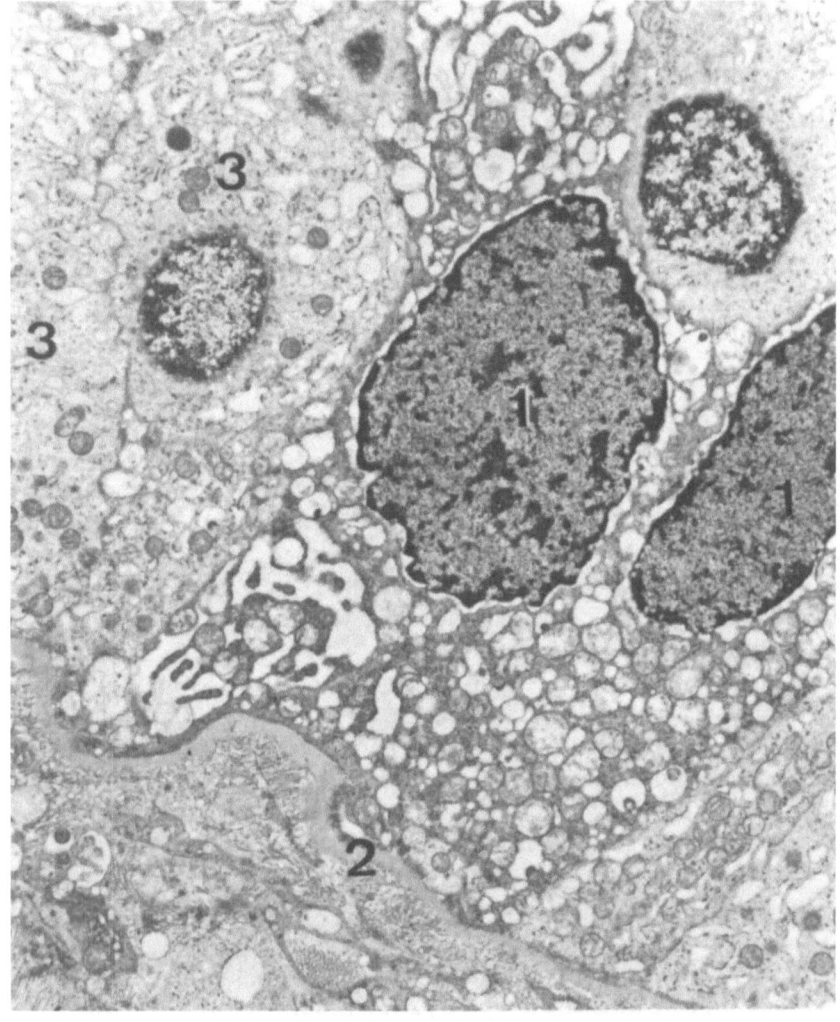

Fig. 15. Degenerating cells in the seminiferous tubule *1*. Basement membrane *2*. S_a Sertoli cells *3*.
X 8,500

merous publications reporting on light microscopic studies of the Sertoli cells (Sertoli, 1865; von Ebner, 1871; Spangaro, 1902; Hoven, 1914; Stieve, 1935; George, 1937; Mancini et al., 1965; Court et al., 1970).

Electron microscopic studies of the Sertoli cells have been carried out mainly on adults (Burgos and Fawcett, 1955; Fawcett and Burgos, 1956; Bawa, 1963; Schmidt, 1964; Nagano, 1966; Sohval et al., 1971; Schulze, 1974). Only isolated reports on Sertoli cells in children are to be found in the literature (Vilar, 1970). According to this author, two types of Sertoli cells can be distinguished in children, namely dark and light. This would correspond to Johansen's (1969) observations in adults.

The Sertoli cell is the most common cell in the infant seminiferous tubule. It is smaller than the spermatogonia and changes its form several times before puberty. Four different types of Sertoli cells can be distinguished in children: fetal (SF), Sa-,

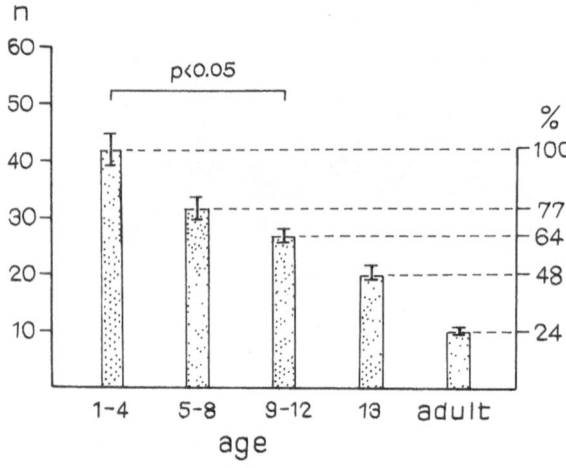

Fig. 16. Decrease in the number of Sertoli cells per cross-section of seminiferous tubule

Sb- and Sc-type cells (Hadžiselimović, 1976). Sa-type Sertoli cells develop from the fetal Sertoli cells in the first year of life. With increasing age, Sb-type cells, which develop from the Sa-type cells, are more frequently encountered. Several transitional stages between these two types can be observed in all age-groups up to puberty. The transition to the Sc-type cells takes place suddenly; with the appearance of the lumen, only Sc-type cells are found in the tubule.

Throughout the entire period of childhood, no division of the Sertoli cells was ever observed: this in a material of 37 normal and 154 cryptorchid testicles in all the age-groups examined. The number of Sertoli cells per cross-section of tubulus decreased continuously in normal testicles from the first year until puberty (Fig. 16). The mean Sertoli cell count in the first four years is 42 per seminiferous tubule, while the figure is only 26 cells per tubule about the age of thirteen.

a) Fetal Sertoli Cells (Fig. 17)

The fetal Sertoli cell is a polarised cell (Hadžiselimović, 1976), its cell-body being divided by the nucleus into a basal and an apical part. The bottom part of the cell lies along the basement membrane and is broader and shorter than the apical part. The nucleus is elliptical, with several invaginations.

There is no cytoplasm halo surrounding the nucleus. The loose reticular nucleolus lies generally in a central position. The mitochondria, which are oval to longish in shape, are of the crista type and are found in groups in both the basal and apical parts of the cell. The Golgi apparatus is well developed, showing both lamellar and vesicular structures, and is generally situated in the vicinity of the nucleus in the apical part of the cell. A typical feature of the fetal Sertoli cells is the parallel and sometimes concentrically arranged rough endoplasmic reticulum, which lies in three to six rows in the cytoplasm. It forms contacts with the lipoid droplets, which occur in the basal as well as the apical part of the cell. In addition to the rough endoplasmic reticulum, smooth endoplasmic reticulum is also found in the cytoplasm of the Sertoli cells. The lipoid droplets can be observed particularly frequently grouped around the lamellar body. This body, formed by the junction of H-shaped lamellae, is found only in the fetal and adult Sertoli cells (Fig. 17), and continues into the rough endoplasmic reticulum. Polyribosomes and glycogen are scattered throughout the cytoplasm. Between

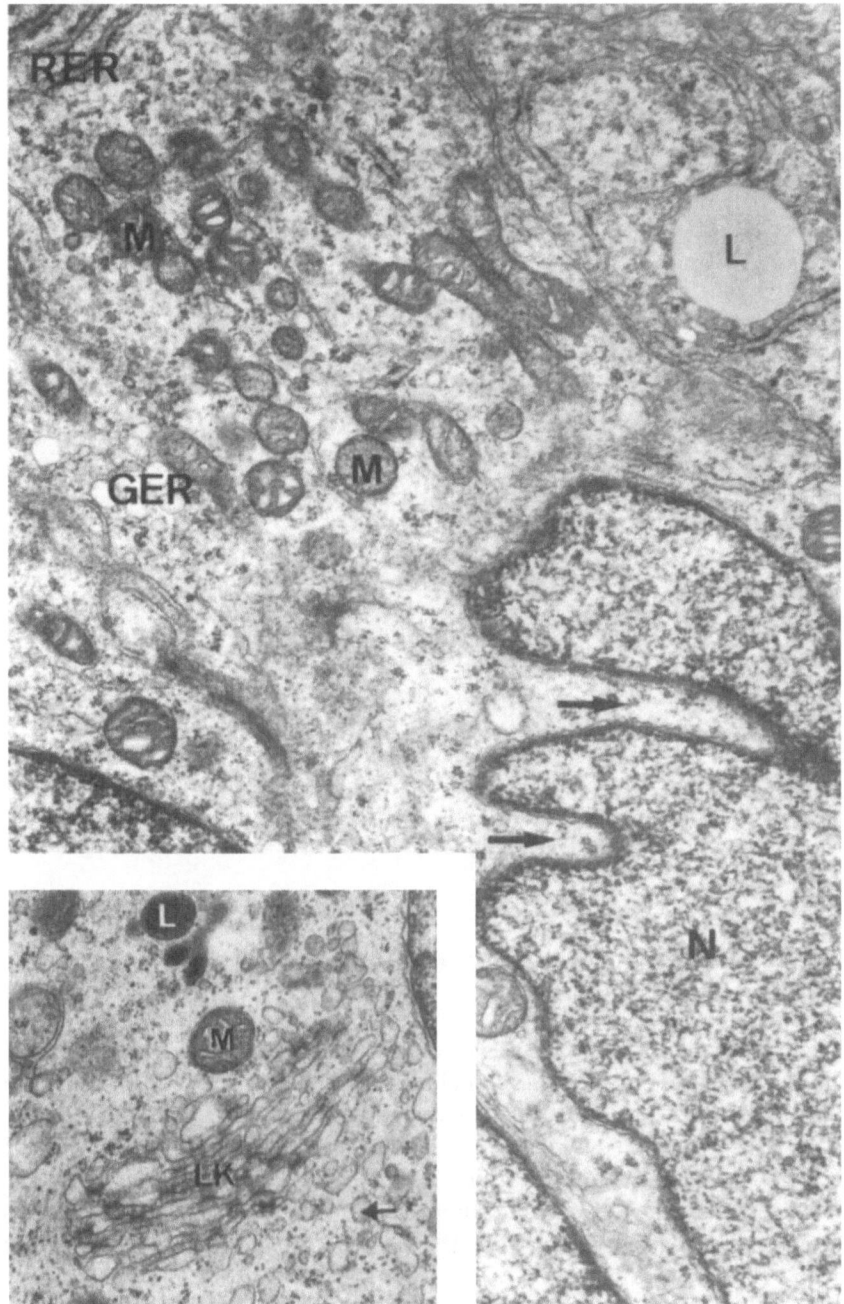

Fig. 17. Fetal Sertoli cells. † shows the nuclear invaginations. *M* mitochondria, *GER* smooth endoplasmic reticulum, *RER* rough endoplasmic reticulum, *L* lipoid. X 17,000.
Enlarged section: Lamellar bodies of a fetal Sertoli cell *Lk, M* mitochondria, *L* lipoid, † indicates
smooth endoplasmic reticulum. X 20,000

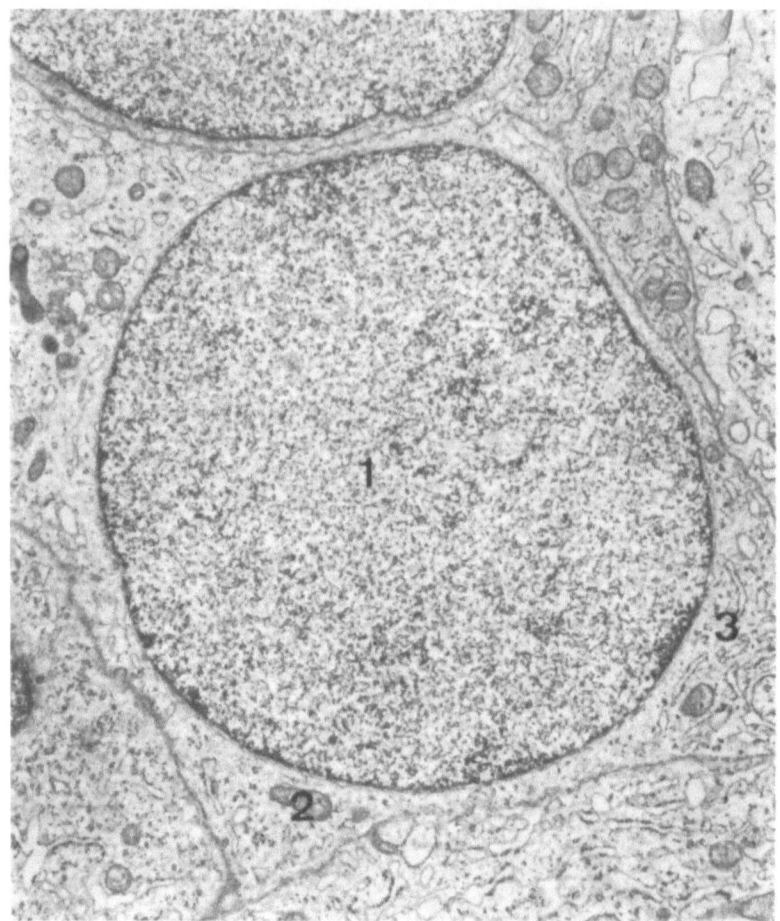

Fig. 18. S$_a$ Sertoli cells. *1* nucleus, *2* mitochondria, *3* rough endoplasmic reticulum. X 10,500

the individual Sertoli cells, complex as well as simple (tight) cell junctions can be observed. In the apical part of the cell, round bodies, which are strongly osmophilic and form clumps, can occasionally be found. Two distinct parts can be distinguished: a round part, centrally situated, strongly osmophilic and reminiscent of the nucleus of a phagocytized cell, and a granular, less osmophilic part lying around the nucleus. The whole body is bounded by a partly interrupted membrane. The degeneration of these bodies, which are reminiscent of phagocytized cells, continues in the Sertoli cell till a clear vacuole forms.

b) Sa-type (Fig. 18)

Immediately after birth, the fetal Sertoli cell changes into the Sa-type Sertoli cell, no more fetal Sertoli cells being found after the third month post natum. The Sa Sertoli cells are roughly twice as big as an adult erythrocyte and are, according to definition, in contact with the basement membrane. No apical or basal cell parts can be distinguished. The entire cytoplasm is arranged concentrically around the round nucleus.

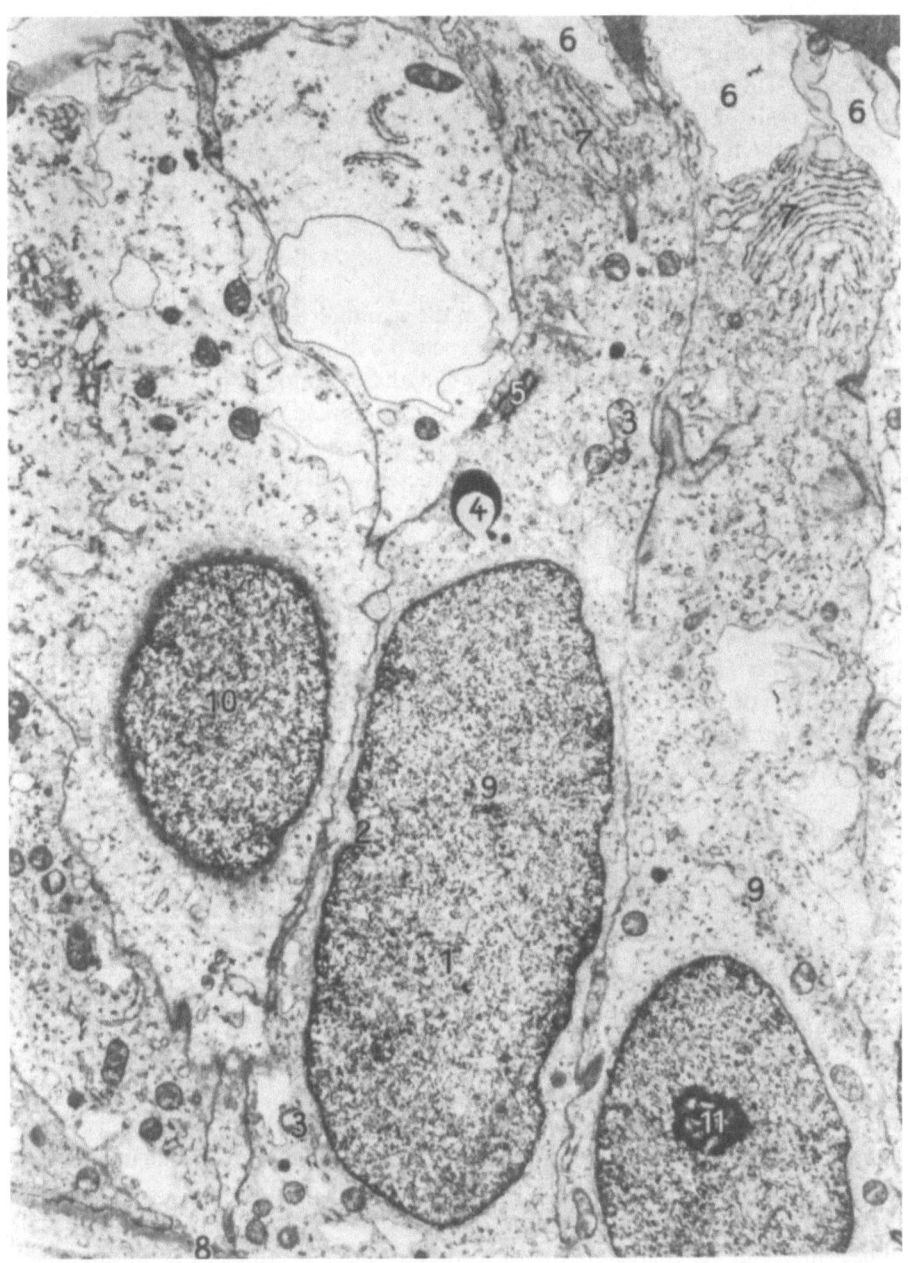

Fig. 19. S_b Sertoli cells. Nucleus *1* with invagination − incipient *2, 3* mitochondria, *4* lipoid, *5* centrosome, *6* vacuoles located typically in the apical part of the cell, *7* rough endoplasmic reticulum, *8* basement membrane, *9* nucleus, *10* transitional cell between S_a und S_b, *11* nucleolus. X 7,000

The nucleoli, which are generally peripherally situated, are not so loosely reticular as those of the fetal Sertoli cell. The homogeneous chromatin is distributed throughout the whole nucleus, which is free from invaginations. The mitochondria are scarce, being of the crista type, with a round to slightly oval body. In the vicinity of the nucleus lies a poorly developed Golgi complex. A small amount of rough endoplasmic reticulum is present, but there is no smooth endoplasmic reticulum and no lamellar body can be discerned. Lipoid droplets appear again only with increasing age. The Sa-type cell remains the most common type of Sertoli cell in the seminiferous tubule until puberty.

c) Sb-type (Fig. 19)

The Sb-type Sertoli cell can be observed in the seminiferous tubule throughout the entire period of childhood. The typical Sb cells are common in the fourth year of life. The cell body is polarised, the nucleus dividing the cell into two parts, with the long axis perpendicular to the single-layer basement membrane. The nuclear membrane is irregular, with one or two and sometimes several invaginations. The loosely reticular nucleolus has a round pars amorpha and is generally situated peripheral to the nucleus. The nuclear membrane is strongly electron-dense. The chromatin forms medium to large clumps, dispersed throughout the nucleus. The mitochondria are round to longish in shape and of the crista type. They are to be found almost equally frequently in the basal and apical parts of the cell. The micropinocytotic vesicles are concentrated in the basal part of the cell membrane. The rough endoplasmic reticulum is situated mainly in the apical part, arranged in three to six rows aligned towards the point of the cell. In the Sb-type of Sertoli cell, smooth endoplasmic reticulum is again found in abundance. With increasing age, the endoplasmic reticulum becomes more differentiate. A characteristic feature of these cells is the emergence of vacuoles in the apical part. These vacuoles are electron-light and have no enclosures. The Golgi complex is well developed and lies in the vicinity of the nucleus. Lamellar and vesicular formations can be distinguished. The lipoid droplets are bounded by a membrane in the cytoplasm and occur regularly from the eighth year onwards. As in the Sa-type cell, no lamellar bodies or crystalloids occur in the cytoplasm of the Sb-type Sertoli cell. The electron density of the Sertoli cell border increases until the age of thirteen. At the point of contact, the two Sertoli cells are often connected by a tight junction.

d) Sc-type (Fig. 20)

With the onset of puberty, and the rise of the gonadotropin level, the Sertoli cells change to Sc-type cells. This change takes place relatively quickly. While in twelve-year-olds, Sa and Sb cells, and Sa-Sb transitional types can still be observed, in thirteen-year-olds all the Sertoli cells are already of the Sc-type. This type of Sertoli cell has almost all the characteristics of the adult cells. According to definition, the Sc-type cell lies on the basement membrane and has a polarised cell-body. The Sc cell is five times bigger than the Sa cell. The nuclear membrane is irregular, with one deep invagination and the chromatin is fine and evenly distributed. The nucleolus is no longer on the periphery, but is centrally situated and consists of loose reticular substance, with one to two round, amorphous bodies. Around the nucleus a light halo of cytoplasm is clearly distinguishable. The mitochondria are smaller than those of the spermatogonia and mainly longish in form. Occasionally boomerang-shaped specimens are also found. All the mitochondria are of the crista type. The Golgi complex is prominent and lies in the vicinity of the nucleus. The rough endoplasmic reticulum is in

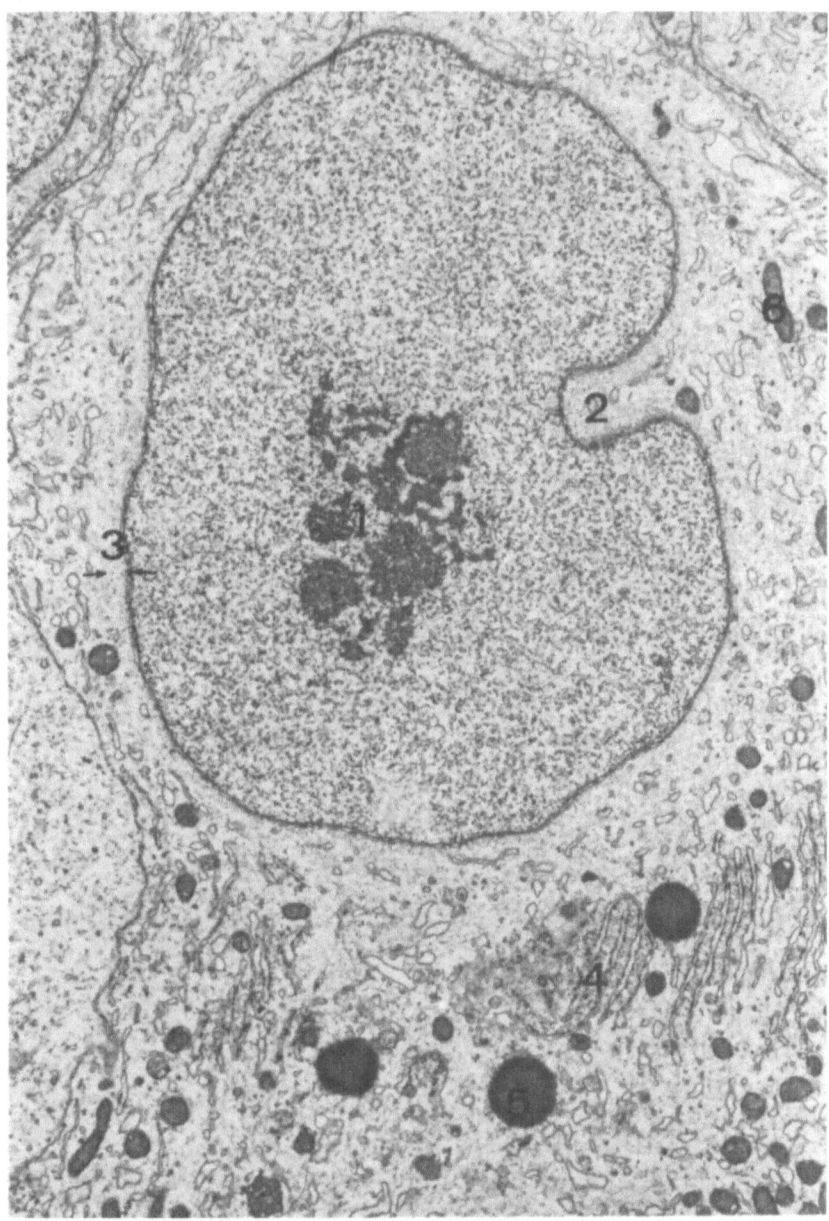

Fig. 20. S$_c$ Sertoli cell. *1* nucleolus, *2* nuclear invagination, *3* light cytoplasm halo, *4* rough endoplasmic reticulum, *5* lipoid, *6* mitochondria. X 12,000

contact with the lipoid droplets. The long, parallel strands of rough endoplasmic reticulum, typical of the fetal Sertoli cells are also to be found in this type of Sertoli cell. The smooth endoplasmic reticulum of the Sc-type cells is the best developed of all the Sertoli cells. Numerous lipoid droplets are distributed throughout the cytoplasm. The four functional phases which Schulze (1974) established in adult Sertoli cells cannot yet be distinguished in the thirteen-year-old.

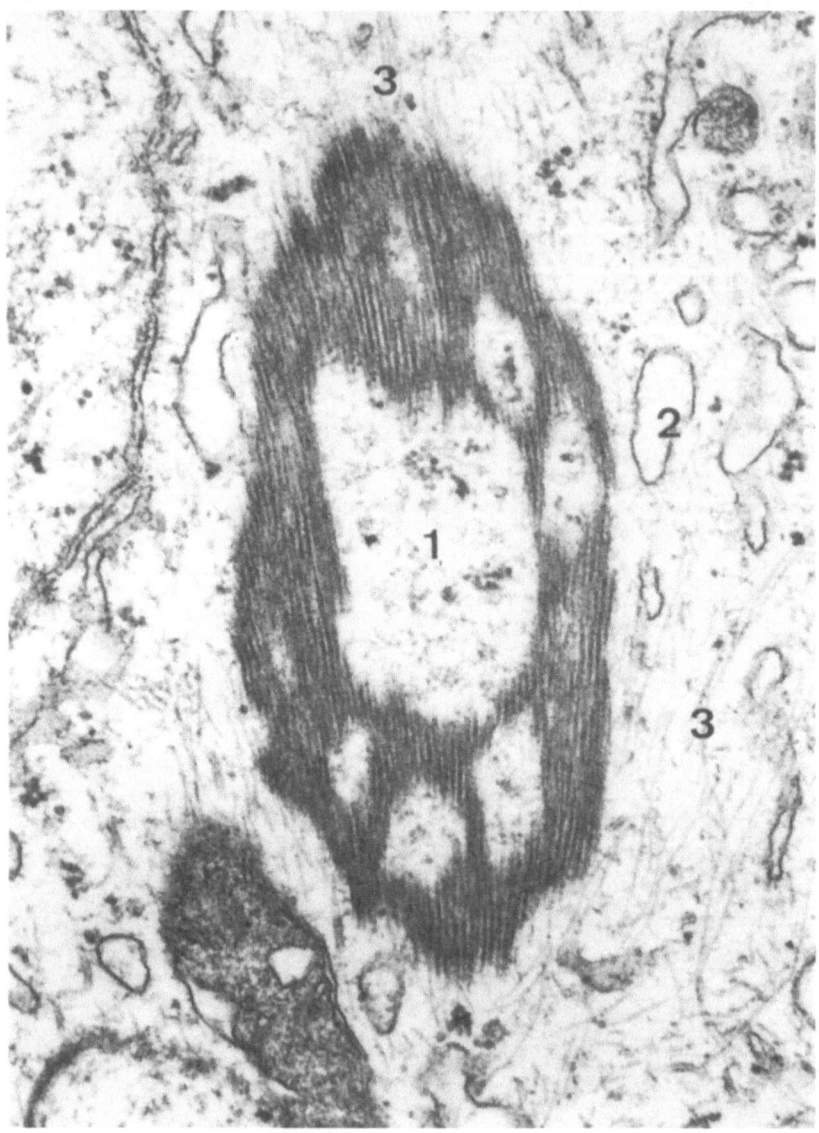

Fig. 21. Charcot-Böttcher's crystalloid. *1* "see-through effect", *2* smooth endoplasmic reticulum, *3* tonofibrils. X 52,000

Charcot-Böttcher's crystalloid (Hadžiselimović and Seguchi, 1974), which is generally situated around the nucleus, is a recently discovered structure in the development of the Sertoli cell (Fig. 21). Only one crystalloid of the Charcot-Böttcher type is present in each cell. It is 5 μ long and 0.8 μ wide and is surrounded by mitochondria, endoplasmic reticulum and frequently also by lipoid granules. In several places fenestrations are visible — the so-called "see-through" effect (Sohval et al., 1966). The crystalloid is composed of approximately 100–150 Å wide fibres, between which there are no electron-dense granules. In the lighter zones, however, granules are clearly visible,

33

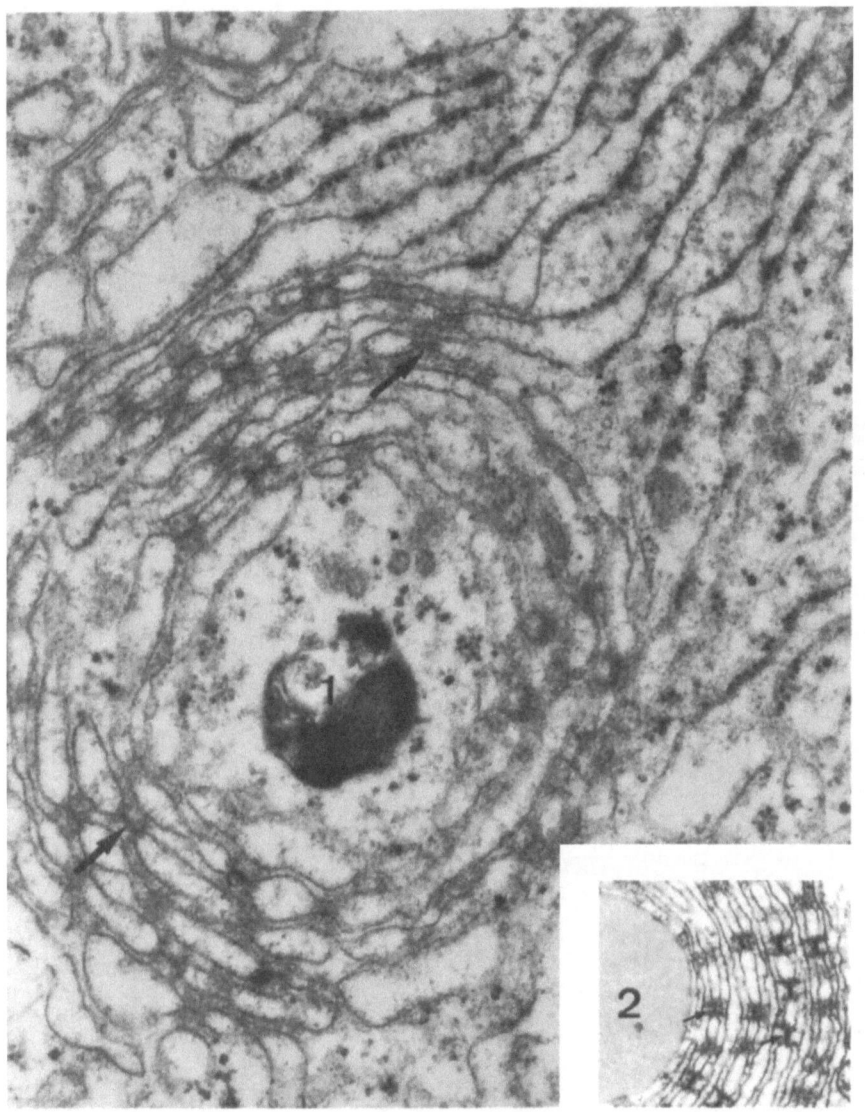

Fig. 22. Formation of the lamellar body from rough endoplasmic reticulum in a 12-year-old cryptorchid boy who was treated with HCG. *1* lysosome-like particles in the centre, *3* rough endoplasmic reticulum, ↑ indicates H-shaped junctions between the individual lamellae. X 44,000.
Enlarged section: Lamellar body of a Sertoli cell of an adult ↑ H-junction, *2* lipoid. X 30,000

which, from their appearance, would seem to be ribosomes. The fibres continue as tonofibrils. There is no limiting membrane round the crystalloid.

The anular membranes in the adult Sertoli cells consist of concentrically arranged double membranes, generally with a lipoid granuloma in the centre (Fig. 22), and surrounded by accumulations of lipoid granules, mitochondria and rough endoplasmic reticulum. The space between the two layers of the double membrane is approximately 320 Å wide and has electron-light contents. H-shaped junctions are to be found at

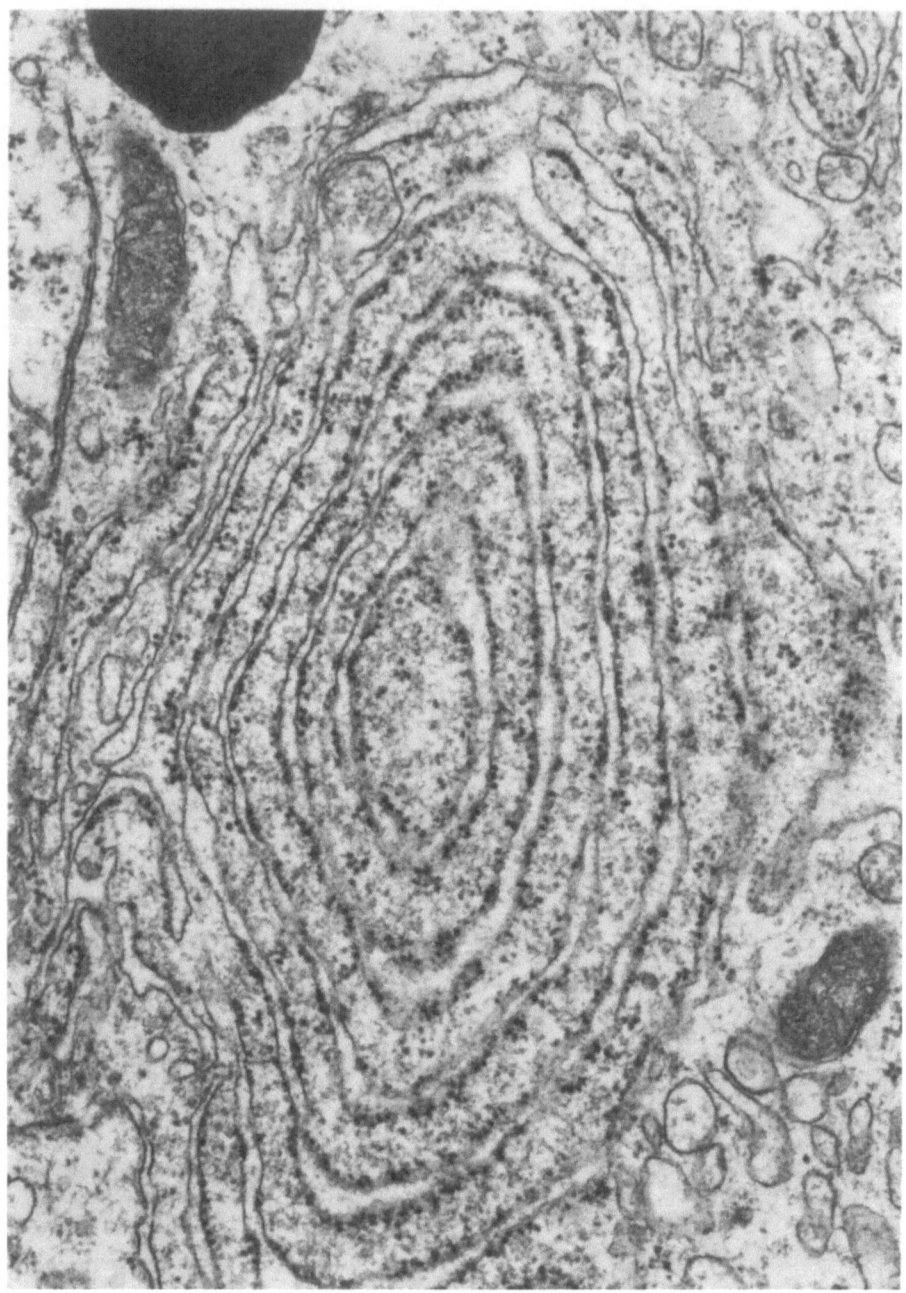

Fig. 23. Precursor of the lamellar body in the cytoplasm of a 13-year-old. X 95,000

intervals of approximately 660 Å in the double membrane. In the transverse sections, these H-junctions appear as pores. In the spaces between the double membranes, the ground substance of the Sertoli cells is visible. Only at the junctions, where the electron-dense substance is deposited, is there no ground substance. Concentric lamellae of rough endoplasmic reticulum can be found in the cytoplasm of thirteen-year-olds

(Fig. 23). Between the individual rings is a relatively large space, in which there are no H-junctions in the centre there are no lipoid granules. At the periphery of these lamellae, the transition from rough to smooth endoplasmic reticulum takes place, while around them lie lipoid granules, mitochondria and rough endoplasmic reticulum. These formations are probably precursors of the anular membrane. After hormone stimulation (for maldescensus), almost completely developed anular bodies are visible in the Sertoli cells of twelve-year-olds (Fig. 22). These differ from the lamellar bodies of adults, having fewer H-junctions and a central lysosome-like formation surrounded by ground cytoplasm in place of the lipoid granuloma found in adults. On the periphery, the anular membranes continue into the rough endoplasmic reticulum (Fig. 22). In the vicinity of these later precursor stages, mitochondria and lipoid granules are also present.

Bawa (1963) and Nagano (1966) described the anular lamellae, e. g. anular bodies in the Sertoli cells of adults. According to them, these lamellae consist of vacuolated smooth endoplasmic reticulum, concentrically arranged around a lipoid granule. Bawa (1963) reports that the fenestration of the anular lamellae is very similar in size to the pore of the nuclear membrane. Kessel (1968, 1973), for his part, describes the formation of anular membranes from elements of the nuclear membrane. He does not however, exclude the possibility of their arising from the rough endoplasmic reticulum (Kessel, 1975). According to our observations, the rough endoplasmic reticulum first forms the anular structure, from which the anular lamellae develop in puberty. Nothing is known about the function of these lamellae. However, the fact that mitochondria have been found in their vicinity, that they are covered with lipoid granules and that they are in close contact with the rough endoplasmic reticulum, gives rise to the hypothesis that they are involved in the production of steroids. The presence of lamellar bodies in fetal as well as in adult Sertoli cells and the appearance of such formations after intensive HCG stimulation, supports the theory that they are dependent on gonadotropin. No mature sperm is found until the fully developed lamellar bodies are present in the cytoplasm of the Sertoli cells.

e) The Function of the Sertoli Cell

From the morphological point of view, certain conclusions may be made about the function of the infant Sertoli cell. Like the adult cell, it has a supporting function in the seminiferous tubule. The primary spermatocytes are nourished by the Sertoli cell, which thus also serves as transport regulator and as part of the blood-testicle barrier, which probably becomes fully developed only in puberty. The fetal Sertoli cell is particularly well adapted for phagocytosis. The gonocytes are actively phagocytized by the Sertoli cells, which pick them up with their pseudopodia and digest them to the vacuole stage (Hadžiselimović, 1976). The presence of the fully developed Sb Sertoli cell seems to have some connection with the appearance of primary spermatocytes and an intensive development of the Leydig cell in the fourth, fifth and sixth year. The polarisation of the cell, appearance of smooth endoplasmic reticulum, lipoid droplets and vacuoles are evidence of the great activity of this cell.

f) Hormone Production

It is not known to what extent the immature Sertoli cells is involved in hormone production. According to several researchers, the adult Sertoli cells are capable of produc-

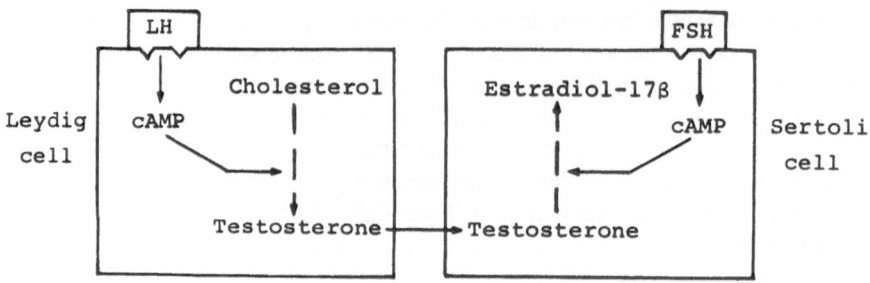

Fig. 24. Model of two cell – two gonadotropin hypotheses for testicular estradiol-17-β secretion. From Dorrington J. H.; Fritz I.B.; Armstrong D.T. In: Regulatory mechanisms of male reproductive physiology (eds. Spilman, C. H.; Lobl, T.J., and Kirton, K.T.)

ing estrogen, Huggins and Moulden (1945). The most recent studies of Dorrington et al. (1976) support the theory that the Sertoli cells have an estrogen-producing role. Their hypothesis is that the Leydig cell, under the influence of LH, secretes testosterone, which then penetrates rapidly into the Sertoli cells, where it is converted by the action of FSH into 17-β-estradiol (Dorrington et al., 1976) (Fig. 24). The Sertoli cell is thus analogous to the granulosa of the ovaries, which have a similar embryological origin (Dorrington et al., 1976). The physiological significance of estrogen synthesis in the testicles is unknown. Dorrington et al. (1976) offer three alternative explanations: firstly, that it is involved with the feed-back regulation of gonadotropin secretion; secondly, that it plays a local role in the seminiferous tubule, by providing suitable conditions for spermatogenesis; and thirdly, that estradiol, which is produced by the Sertoli cell, plays some local regulatory role in the metabolism of the Leydig cell.

In all probability, the Sa Sertoli cell plays no part in hormone production. To what extent Sb Sertoli cells are involved in secretion is not certain, although a morphological analogue does exist. It may be assumed, however, that fetal and Sc Sertoli cells do have a hormone-producing role, evidence of this being provided by the numerous lipoid droplets, the abundance of smooth endoplasmic reticulum, the large number of osmophilic and the lamellar body in the cytoplasm. According to Josso (1974), the fetal Sertoli cells produce the so-called "Factor X", which has a Müllerian-inhibiting activity. A morphological analoque of this exists.

4. Peritubular Connective Tissue

The peritubular connective tissue can be subdivided into the basement membrane and tunica propria, which in humans consists of an inner collagen fibre zone and an outer cellular zone (Yusawa, 1968). In the inner zone, Schmidt (1964) observed sparse collagen fibres, with their long axis parallel to that of the tubulus. Yusawa (1968) distinguishes three layers in the peritubular connective tissue: basement membrane, collagen fibre zone and a cellular zone composed of fibroblasts. He observed a basement membrane consisting of one or several layers, with single lamellae 0.1 μ thick. In the cellular zone he found two different types of fibroblasts. According to his observations, there are no muscular elements in the tunica propria. However, Clermont (1960), Lacy and Rotblat (1960), Lesson (1963) in the rat, Baillie (1964), Gardner and Holyoke

(1964), Ross (1967) in the mouse and Rothwell and Tingari (1973) in domestic fowl report a type of contractile cell in the peritubular structure which is electronmicroscopically similar to the smooth muscle cells.

Fawcett and Burgos (1960), Ross and Lang (1966) and Mazzuca (1971) have described muscle-like elements in the cellular zone of the peritubular structure in man. Roosen-Rungen, (1951), in observations in rats and dogs, was actually able to establish that the seminiferous tubule has the ability to contract.

a) The Development of the Peritubular Structures

In children, the basement membrane and the tunica propria can be clearly delimited in the peritubular zone. The tunica propria is further subdivided into two zones: an inner zone, which contains collagen fibres and an outer zone consisting of fibroblasts. The peritubular zone gradually gives way to the interstitium. This transitional zone is relatively well developed and may almost be designated as a separate zone. Since, however, its collagen fibre structure has no definite zonal boundary and becomes lost, particularly in the interstitium, it is better to refer to it as a transitional zone (Seguchi and Hadžiselimović, 1974).

There is a marked difference not only in the appearance, but also in the composition of the peritubular connective tissue in one- and thirteen-year-olds. These differences become clear on direct comparison of these two age-groups.

b) One-year-olds

According to definition, three zones can be distinguished in the one-year-old peritubular connective tissue (Fig. 25). The zone nearest the tubulus is the basement membrane, which is composed of two layers, the inner layer being thin and electron-light. No structures are present in this layer. The second, outer layer is homogeneous, electron-dense and thicker than the inner layer, the width remaining constant throughout its extent. In one-year-olds small, differentiated, knob-like formations are rarely found on the basement membrane. The relatively narrow second zone consists of irregularly arranged collagen fibres, which are relatively scarce here, compared with thirteen-year-olds: some parts are even completely free of collagen fibres, in which case the zone is very narrow. The third zone consists of one or two fibroblast layers, in which two distinct types of fibroblasts can be seen. Those lying close to the tubule have an oval nucleus, rich in strongly electron-dense chromatin and generally located on the nuclear periphery. One to two undeveloped nucleoli are to be found, mostly situated peripherally and in contact with the nuclear membrane. In the electron-light cytoplasm, round mitochondria of the crista type can be distinguished. The rough endoplasmic reticulum is tubular and situated in the processes. Ribosomes and free glycogen bodies are dispersed throughout the cell, which has an undulating surface, with occasional micropinocytotic vesicles. All the cells are arranged concentrically around the tubulus.

The second type of fibroblast is also situated on the periphery of the cellular zone (Fig. 26). This second type differs from the first mainly in the appearance of the nucleus, which is more elongated, with numerous deep invaginations on the surface. The cytoplasm is electron-dense, occasionally with a few lipoid droplets. The endoplasmic reticulum, particularly the rough, seems to be more abundant. These cells can be considered as precursors of the Leydig cells (Fawcett and Burgos, 1960). Plasma bodies

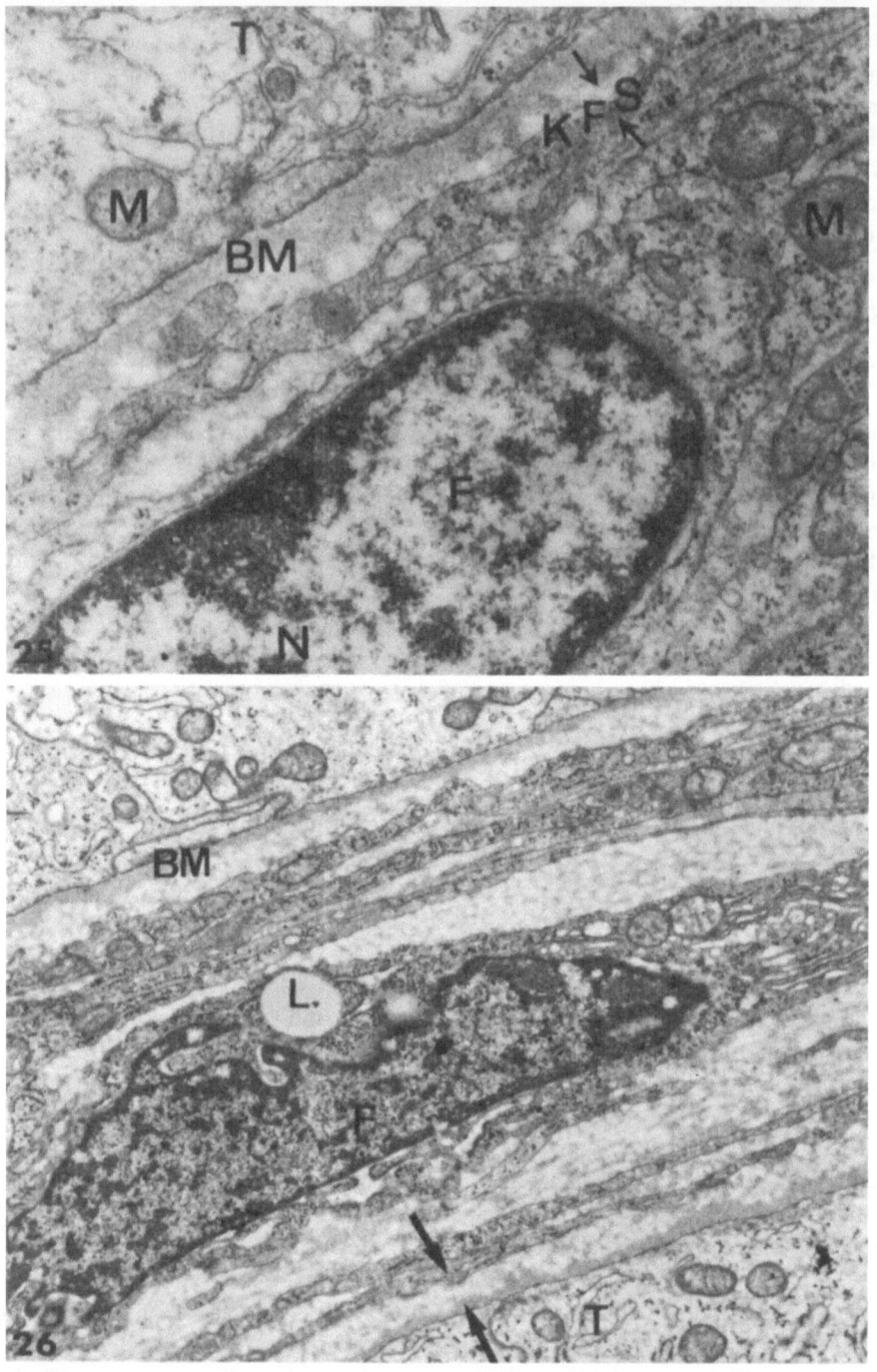

Fig. 25. Peritubular connective tissue of a 1-year-old boy. The zone apposed to the tubulus is the basement membrane *BM*. Relatively narrow collagen fibre layer *KFS*. The third zone is composed of fibroblasts *F*, nucleus *N*, mitochondria *M*. X 19,500

Fig. 26. Peritubular connective tissue of a 2-year-old boy. *T* seminiferous tubule, *BM* basement membrane. The type II fibroblast which corresponds to the precursors of the Leydig cells *F* has an irregular nucleus and lipoid droplets in the cytoplasm *L*. X 15,000

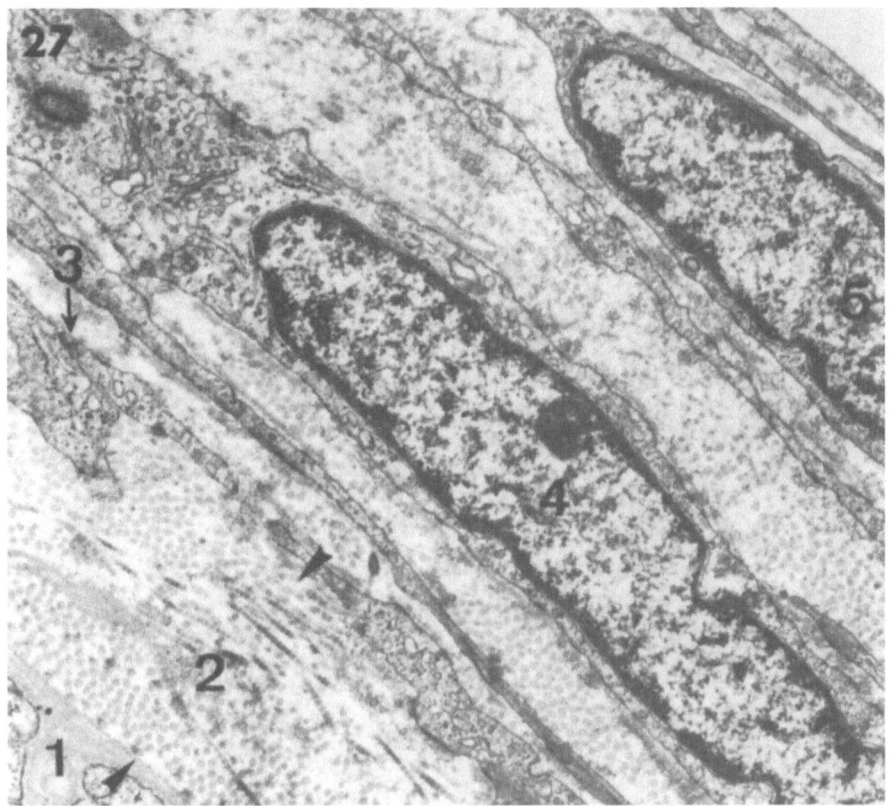

Fig. 27. Peritubular connective tissue of a 13-year-old. The basement membrane is layered, with well-developed knob formations *1*. The collagen fibre layer is broad *2*. The cellular zone consists of myofibroblasts *4* and fibroblasts *5*. Tight junction between two myofibroblasts *3*. X 15,000

also occur here and there in the cellular zone. Since the interstitium of one-year-olds is very tightly packed with Leydig cells and their remnants, continuous transitions are rarely found.

From the age of one to thirteen, i. e. until puberty, the changes occurring in the peritubular connective tissue are quantitative rather than qualitative. The basement membrane retains its two-layered structure until puberty. The collagen fibre layer increases in thickness and width, but no organised development of the collagen fibres takes place until the age of twelve. The cellular zone consists of fibroblasts until puberty, the number of layers generally varying from three to six. Between the cells of the cellular zone, precursors of the collagen fibres can be detected. These topocollagen islets generally become clearly visible in the ninth and tenth year.

c) Thirteen-year-olds

The peritubular connective tissue in thirteen-year-olds is composed of the basement membrane and tunica propria, the latter being here clearly distinguishable from the interstitium. The basement membrane is split up into several layers: approximately ten layers can be counted. Close to the tubule lies an electron-light layer, followed

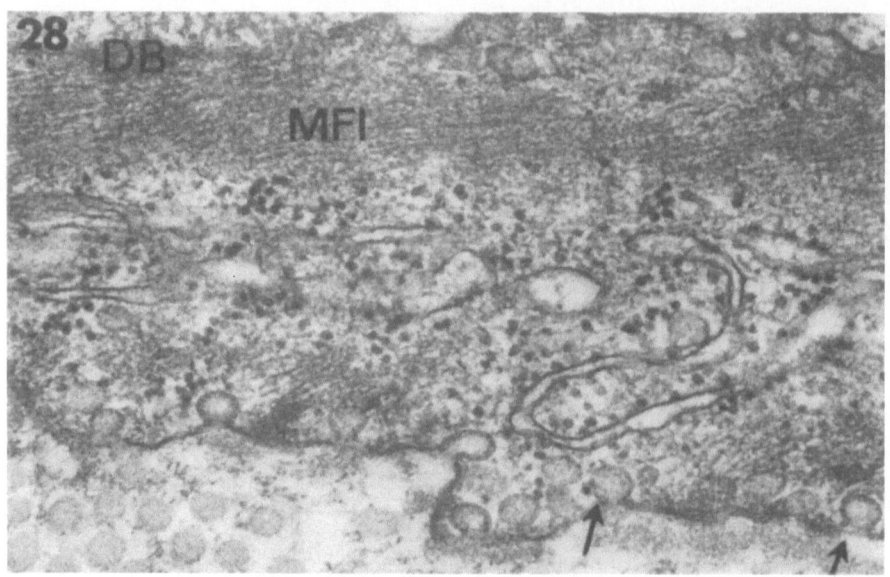

Fig. 28. Enlargement from the cytoplasm of a myofibroblast. ↑ micropinocytosis, *MFI* myo-
filaments, *DB* dense bodies. X 36,000

by alternate electron-dense and electron-light layers. The width of the membrane varies considerably. With the onset of puberty, the knob on the basement membrane also
becomes clearly visible. It is in contact with the Sertoli cell only and is very often the
point of junction of several Sertoli cells. The basement membrane is followed by a
broad collagen fibre layer, the fibres of which are aligned in a circular or longitudinal
direction. The fibroblast zone, which follows upon the collagen fibre zone, consists
of three to six layers, with the cells arranged concentrically around the tubule. Two
different cell-types occur, the type nearer to the tubule being typical of the tunica
propria of the tubuli from the age of thirteen onwards. They have long processes,
narrow and with electron-dense cytoplasm. The elliptical nucleus is situated in the
middle of the cell. The nuclear membrane is very irregular, with deep invaginations.
The nucleolus is loose reticular and generally situated on the periphery. In the cytoplasm, long mitochondria of the crista type are found. The rough endoplasmic reticulum is well developed, tubular in form and is in contact with the mitochondria and
the nuclear membrane (Fig. 28).

The characteristic features of this cell are:
1. The filaments lying on the periphery of the cell.
2. Extremely active micropinocytosis.
3. Dense bodies, mostly in contact with the cell membrane.
4. Tight junctions with other cell processes of the same type.

Since there was no migration of muscle elements into the peritubular structure during the development of the testicle in children, it would thus appear that these cells
develop from the fibroblasts and consequently should be classified as so-called "myofibroblasts" (Majno et al., 1972).

Fibroblasts are found in this cellular zone, towards the interstitium (Fig. 28). These cells are also arranged concentrically around the tubulus and have long, narrow processes, the cytoplasm of which contains very few cell organelles. Micropinocytosis is rare. The characteristic muscle structures mentioned above do not occur here. The cell nucleus is elliptical and undulating, with no deep invaginations, very little chromatin and one to two nucleoli. The cellular zone continues into the interstitium, in the form of a field of collagen fibres. Until the thirteenth year, neither elastic fibres nor reticular fibres are found.

As already mentioned, the basement membrane is two-layered until the thirteenth year and only in puberty does it become multilayered. It was possible to show that this process is related to gonadotropin, thus indicating that the peritubular connective tissue, like the germinal epithelium and the transitional Leydig cells, comes to final maturity in puberty, under the influence of gonadotropin. From the narrow, partly inhomogeneous basement membrane develops a multilayered band about 800 Å wide, directly apposed to the germinal epithelium. (Seguchi and Hadžiselimović, 1974; Burgos, 1960; Altdorfer and Hedinger, 1977). The knob formation also seems to be under the influence of gonadotropin. Immediately after birth, when the gonadotropin level is still high, a rudimentary knob can occasionally be found on the basement membrane. At the age of nine, i. e. at the beginning of puberty, the knob formation also becomes visible and well developed, lamellar knobs can be seen in the thirteenth year.

d) The Function of the Peritubular Connective Tissue

The peritubular connective tissue fulfils several functions. In addition to its supporting role, it also *has the capacity to produce periodic contractions* (Roosen-Runge, 1951). In 1958, Clermont was able to demonstrate the periodicity of the contractions on isolated tubuli of the rat testicle. The contractions occur at intervals of thirty to forty-five seconds, the tubular wall contracting with an amplitude of up to 10 μ. The contractions seem to move in the direction of the rete testis. Also the isolated tunica propria, i. e. without the tubulus seminiferus contents, has the capacity to contract (Clermont, 1960). Setschell studied in detail a further function of the peritubular structure, namely that of metabolic transport regulator. He distinguished an active and a passive transport, dividing substances into three groups, according to flow rate. The first group includes substances which rapidly pass the blood-testis barrier, such as water, urea, ethanol and bicarbonate. In the second group are those substances which rapidly penetrate into the testicular lymph system, but slowly into the rete testis, such as creatinin, galactose, sodium, potassium, chlorine and iodine. The third group comprises substances which penetrate rapidly into the testicular lymphatic ducts, but which cannot be traced in the rete fluid. This group includes, inter alia, the albumina, glutaminic acid, para-aminohippuric acid and inulin (Setschell, 1970). According to Fawcett et al., (1970), the blood-testicular barrier is located not only in the peritubular structure, but also in the Sertoli cells themselves.

Like all other elements of the testicle, the tunica propria is also under the influence of hormones. Altdorfer and Hedinger (1977) believe that the peritubular connective tissue displays distinctly different behaviour in various diseases of the testicles which are due to hormonal influence or the result of reactive hormone changes. The basement membrane can show massive thickening, without any corresponding reduction in the diameter of the tubulus. The number of collagen fibres can also be greatly increased, either almost exclusively in the innermost layer, directly under the basement membrane (reduced gonadotropin) or almost exclusively in the outermost layers (increased gonadotropin, Klinefelter syndrome).

5. Leydig Cells

In a publication entitled: "Zur Anatomie der menschlichen Geschlechtsorgane und Analdrüsen der Säugetiere", Leydig (1850) described for the first time the interstitial cells which have since been known as Leydig cells. Fifty years after the discovery of the Leydig cells, Bouin and Ancel (1903) ascribed an internal secretory significance to them, namely the secretion of the male sex hormone. Ito and Oinuma (1939) explained the functional significance of human testicular interstitial cells from the morphological standpoint. They described the Leydig cell as a cell with many different forms, frequently with long, narrow processes, which often formed a connection between neighbouring interstitial cells. The round or oval nucleus of the interstitial cell is always situated very excentrically, i. e. in one pole of the cell-body. These authors were also able to demonstrate morphologically the secretory processes, which are similar to those of the exocrine gland cells. This led them to the conclusion that the interstitial cells must be the endocrine gland cells. The mitochondria of the interstitial cells, i. e. their number, is in inverse proportion to the number of secretory granules and vacuoles, varying according to the functional condition of the cells. Pigment granules and lipoid in the Leydig interstitial cells have an incretory significance, participating in hormone production (Ito and Oinuma, 1939).

Detailed light-microscopic descriptions of the adult Leydig cells have been given by Stieve (1930), Rasmussen (1932), and Sniffen (1950). However, a functional dynamic analysis of the morphological picture was possible only with the advent of electron microscopy and cytochemistry.

a) The Development of the Leydig Cells

The Leydig cells originate from mesenchymal cells, which are capable of changing into fibroblasts as well as into precursor Leydig cells. The ultrastructure of the fetal Leydig cell was described in detail by Niemi and Pelliniemi (1969), Gondos and Hobel (1971) and Holstein et al. (1971). According to Holstein et al., human embryos in the eighth week of pregnancy have Leydig cells, which have differentiated from mesenchymal cells. The marked increase in size of the cells between the ninth and twelfth week of pregnancy is occasioned by the development of a smooth, vesicular endoplasmic reticulum, an increase in the size of the Golgi complex and development of the cisternae containing ribosomes and numerous cytosomes. In the twelfth week of pregnancy, two types of Leydig cells can be distinguished on the basis of different stages of differentiation of the endoplasmic reticulum and other morphological details. The endoplasmic reticulum attains a high degree of differentiation, in the form of straight tubules, passing through the cell-body in bundles. After the fourteenth week of pregnancy, many Leydig cells show degenerative changes and some die off. The development of Leydig cells precedes the germ cell development by approximately two weeks and reaches its maximum between the twelfth and fourteenth week of pregnancy. After the seventeenth week of pregnancy, the relative and absolute volume of the Leydig cell diminishes (Holstein et al., 1975).

Only a few electron-microscopic data exist about the postnatal Leydig cell (Vilar, 1970; Hadžiselimović and Seguchi, 1975). Light-microscopic studies of the postnatal development of the Leydig cells have been carried out by Mott, 1919; Diamantopoulus, 1921; Lahm, 1922; Sniffen, 1950; Albert et al., 1953; and Mancini et al., 1963). According to these authors, the Leydig cells develop only after puberty. From the end of the first year post natum until puberty, no Leydig cells were visible under the light microscope. Hayashi and Harrison (1971) have studied in detail the development of the Leydig cells from immediately after birth until puberty. In contrast to the above researchers, they were able to distinguish well developed Leydig cells in one-year-olds. After this time, they disappear, to reappear between the age of five and six. From this period onwards until puberty, the Leydig cells increase continuously in number and size. (Hayashi and Harrison, 1971).

As already mentioned on page 38, Leydig cells and their precursors were visible in the interstitium throughout childhood. Precursor Leydig cells are generally fusiform, with a few atypical cell organelles. Two regressive phases in the development of the Leydig cells were observed: the first in the second and third year and the second from the age of nine to eleven. In this period, the fully developed Leydig cells are almost non-existent, but not their precursors. After the age of thirteen, the Leydig cells become more numerous and more differentiate.

b) The First Year

In newborns, the Leydig cells lie in groups of two to three in the interstitium, generally in the neighbourhood of the vessels. Their surface is polygonal and covered with microvilli (Fig. 29). The large, round nucleus is situated excentrically in the cytoplasm, with a loose reticular nucleolus generally visible on the nuclear periphery. It has no contact with the nuclear membrane. The lipoid droplets are visible in groups, even under low magnification. They have a diameter of approximately 1 μ und their contents are mostly dissolved by the methods of preparation. They often have contact with mitochondria, 1 to 1 1/2 μ in size and of the crista type. They are to be found throughout the cytoplasm. The typical feature of the Leydig cell, its smooth endoplasmic reticulum, occurs in vesicular form and occupies the greater part of the cytoplasm of these cells. The rough endoplasmic reticulum is often to be found at the cell periphery. In the vicinity of the cell boundary and the endoplasmic reticulum, membrane-bounded bodies with homogeneous contents are visible. These bodies, which are very reminiscent of microbodies, are mostly round, with a diameter of 2,800 Å (Fawcett and Burgos, 1960). On the surface of the cell, micropinocytotic vesicles are found. Free ribosomes and glycogen granules are distributed throughout the cell. The Leydig cells are joined to each other by tight junctions. Reinecke's crystalloid was not observed in the cytoplasm of neonatal Leydig cells. Compared with the adult Leydig cell, the Golgi apparatus is relatively poorly developed. It is generally situated in the vicinity of the nucleus and possesses lamellar as well as vesicular structures.

c) Leydig Cells in Six-year-old Boys (Fig. 31)

The Leydig cells in the interstitium of a six-year-old boy are representative of the period between the fifth and eight year. They are found in groups of two to six cells in

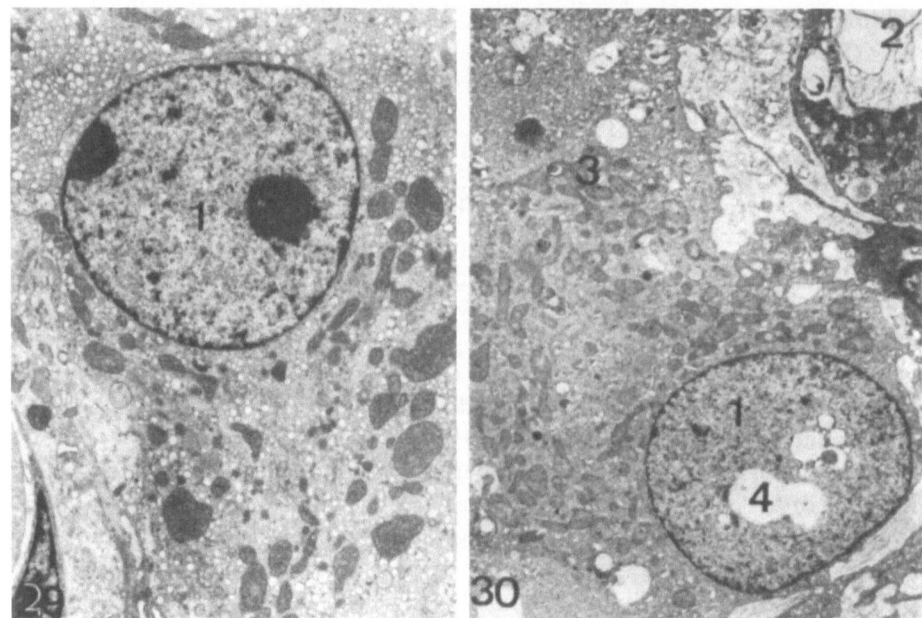

Fig. 29. Leydig cell of a 5-week-old normal testicle. *1* Nucleus. X 4,400

Fig. 30. Atrophic Leydig cell of a 2-week-old cryptorchid testicle. The nucleus *1* has fenestrations *4*. *3* mitochondria. Nearby is a degenerating Leydig cell *2* with vacuoles in the cytoplasm. X 4,400

the interstitium. On the whole, they are smaller than the neonatal Leydig cells, with a polygonal appearance and microvilli covering the cell surface. The chromatin is homogeneously distributed, while the heterochromatin is to be found on the nuclear periphery. Unlike the neonatal Leydig cell, the nucleolus in the six-year-old cell is very well developed, with an easily distinguishable pars amorpha and pars reticularis. The nucleolus is peripherally situated and in contact with the nuclear membrane. The smooth endoplasmic reticulum occurs in vesicular form and fills almost the entire cell. A few mitochondria of the crista type are visible. The lipoid droplet content is much reduced. Ribosomes and glycogen granules are occasionally visible in the cytoplasm. Very rarely microbodies can be seen, generally on the cell periphery. No Reinecke's crystalloid is discernible.

d) Puberty (Fig. 32)

With the onset of puberty, the Leydig cells become bigger and more common in the interstitium. The cytoplasm develops and the nucleus assumes the round form. It is large, with a prominent, loose reticular nucleolus. The cytoplasm again contains lipoid droplets, whose contents have been partially extracted. The mitochondria are oval and of the crista type, with smooth endoplasmic reticulum, in vesicular form, lying between them.

The rough endoplasmic reticulum is not so common as in the neonatal Leydig cells. It lies on the cell periphery and forms short, parallel tubes. The Golgi apparatus is composed of parallel lamellae and vesicles. It is large and found throughout the cytoplasm.

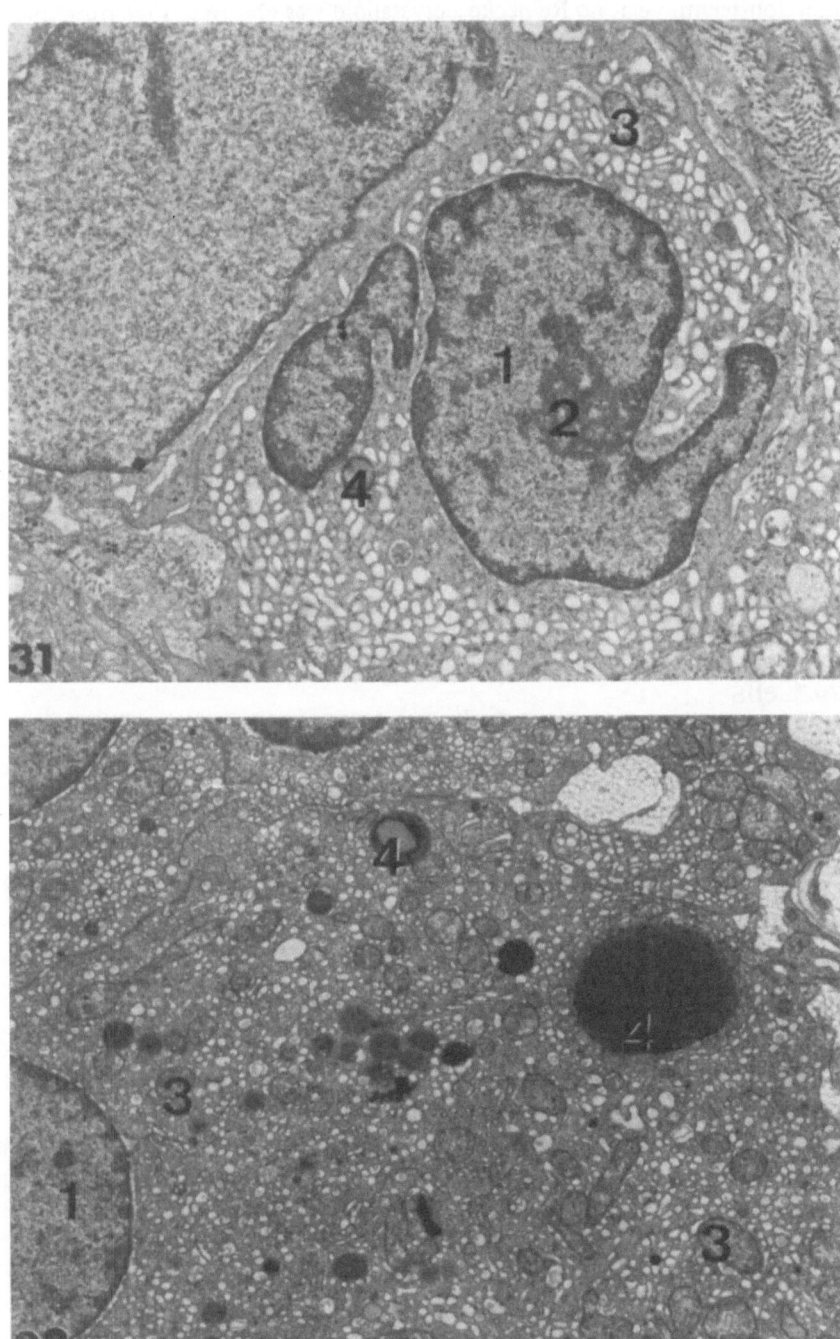

Fig. 31. Leydig cell of a 6-year-old normal testicle. *1* nucleus, *2* nucleolus, *3* mitochondria, *4* lipoid. X 29,000

Fig. 32. Detail of the cytoplasm of a 13-year-old. *1* nucleus, *3* mitochondria, *4* lipoid. X 9,700

Until the fourteenth year, no Reinecke's crystalloid was observed. Free ribosomes lie between the tubules but are not in contact with the smooth endoplasmic reticulum. The Leydig cells are not bounded by the basement membrane.

The second type of Leydig cell, which was observed throughout the whole period of childhood, is the so-called "precursor" Leydig cell (Fawcett and Burgos, 1960) (Fig. 26). These precursors are fusiform. Their nucleus is longish, with several invaginations and peripherally situated chromatin. A loose reticular nucleolus can be observed on the periphery of the nucleus. The cytoplasm has a small Golgi apparatus and few mitochondria of the crista type. Poorly developed rough endoplasmic reticulum is found, particularly in the cell processes. Free glycogen and ribosomes are dispersed throughout the cytoplasm. In the vicinity of the nucleus lie isolated lipoid droplets. The cell membrane exhibits micropinocytotic vesicles. The ground cytoplasm has fine filaments less than 50 Å in diameter and lying parallel to each other.

Cryptorchidism

1. Germ Cells

In research into cryptorchidism, particular attention has always been paid to the germ cells. They were the most important parameter in assessing the quality of the tissue (Hedinger, 1971; Hecker, 1971; Bodensky and Regele, 1973; Jendricke et al., 1973). According to these authors, the number of spermatogonia per tubule in cryptorchid boys is almost the same as in the controls. These results would appear to indicate that the decrease in the number of spermatogonia is a secondary phenomenon, resulting from the unfavourable situation of the testicle. Many researchers, including Scorrer and Farrington (1971), Städtler and Hartmann (1972) and most recently Hedinger (1976) found that the number of spermatogonia in cryptorchid children at birth was already reduced in comparison to normal control testicles. In 1976, Hedinger repeated his studies on 619 biopsies from 415 boys from birth up to the age of ten, with unilateral or bilateral cryptorchidism and established that already in the first two years the number of spermatogonia in the cryptorchid testicles was lower than in the testicles of normal control subjects. From the age of three, cell movement in the cryptorchid testicles remains at about the same low average level, while normal testicles from the same age-group show an increase in the number of spermatogonia. This is also, at least to some extent, demonstrable in the descended testicle in cases of unilateral cryptorchidism, with normal values being recorded only in some cases. In more than half the cases, however, there is a marked reduction in the values recorded, with a marked falling off after the age of six. Occasionally the figures may be as low as those in cryptorchid testicles (Hedinger, 1976).

Hecker et al. (1976) obtained similar results in the descended testicle in cases of unilateral cryptorchidism. According to these authors, the contralateral testicle had a reduced number of spermatogonia in 52.7 % of cases; 32.6 % had a normal spermatogonia count and 14.7 % had no spermatogonia. On the basis of experimental studies carried out by Shirai et al. (1966), Hecker (1971) postulates that the changes in the contralateral descended testicle are of a secondary nature.

Shirai et al. (1966), in experiments with dogs, were able to show conclusively that when one testicle is artifically situated in the inguina, pathological changes occur also in the other, descended testicle. This would indicate that, in cases of artificial crypt-orchidism, the cryptorchid testicle damages the other testicle by some means as yet unknown: an immunoreactive or neural process would appear to be responsible. Until the present, however, the question whether the spermatogonia are genetically damaged from the outset, or whether they disappear from the seminiferous tubule as a result of the unfavourable position of the testicle remains unanswered.

Our own electron microscopic studies were carried out on 154 biopsies taken from cryptorchid children between the ages of two weeks and fifteen years. Of these 154 biopsies, nine, obtained from unilateral cryptorchid children with no other illnesses and no signs of malformation in the region of the testis and epidydimis, were used for morphometric studies. Six biopsies from normal testicles served as controls. The crypt-orchid group consisted of two biopsies each from one-year-old, two-year-old, three-year-old, five-year-old and one biopsy from four-year-old testicles. The control group of normal testicles comprised one biopsy each from one-year-old, three-year-old, four-year-old and five-year-old and two biopsies from two-year-old testicles (Hadziselimovic et al., 1975). The volume density of the spermatogonia was examined morphometrical-ly and was found to amount to 0.056/ccm testicular tissue in the control testicles and 0.021 spermatogonia/ccm testicular tissue in the cryptorchid testicles, which is equiv-alent to a decrease in volume density of spermatogonia of approximately 60 %. The difference between these two groups is thus significant, with P < 0.005 (Table 2).

a) The First Year

In the first year, in addition to the fetal spermatogonia, gonocytes are also occasional-ly present in the seminiferous tubule of cryptorchid infants. The fetal spermatogonia are ultrastructurally identical to those in normal testicles in children of the same age. Rarely, transitional and Ap and Ad spermatogonia can also be observed. The Ap and Ad and transitional spermatogonia appear to be fewer in number than in the normal control testicles.

b) The Third Year

By the age of three, the qualitative changes in the spermatogonia of cryptorchid tes-ticles have become marked. Binuclear spermatogonia are encountered more frequently than in normal testicles (Fig. 33). The spermatogonia are mostly fetal, sometimes with bizarre nuclear structures. The mitochondria of these fetal spermatogonia, which are reduced in number, are of the crista and tubule type. In contrast to normal spermato-gonia, they have an increased number of large, membrane-bounded vacuoles, with in-homogeneously distributed contents (Fig. 33). These could be phagocytized cell ele-ments, although no active spermatogonia phagocytosis was observed. Histometric com-parison of the spermatogonia of three-year-old and one-year-old cryptorchids shows a considerable alteration in the nucleus-plasma ratio in favour of the nucleus (Lüdin, 1977), the cytoplasm of most spermatogonia appearing electron-darker than normal spermatogonia of the same age. The Golgi apparatus and the endoplasmic reticulum show scarcely any development.

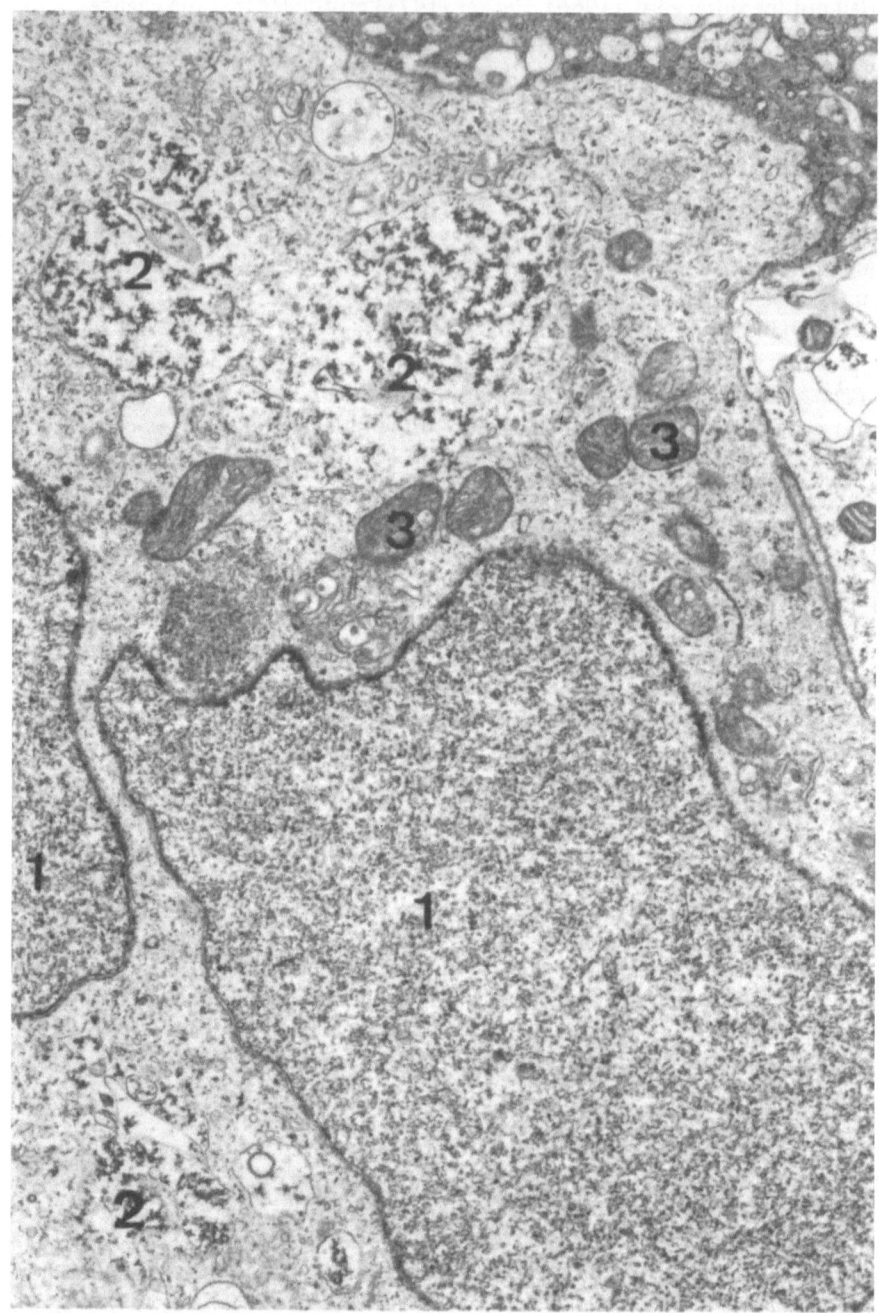

Fig. 33. Binuclear spermatogonium of a 3-year-old cryptorchid testicle. *1* nucleus, *2* vacuoles, *3* mitochondria. X 12,500

c) The Sixth Year until Puberty

The marked reduction in spermatogonia in the six-year-old is obvious: fetal, transitional, Ap and, rarely, Ad spermatogonia are visible. B spermatogonia and primary

spermatocytes are never seen. All the forms of spermatogonia have a narrower cytoplasm, which appears electron-darker than that of the control spermatogonia, and which becomes increasingly reduced with age. The excentrically situated nucleus has a loose reticular, well developed nucleolus, with round mitochondria, which vary greatly in size. Large, membrane-bounded vacuoles, sometimes in contact with mitochondria, are also frequently present in the cytoplasm of all types of spermatogonia. The Golgi apparatus and endoplasmic reticulum are poorly developed. The majority of the spermatogonia show signs of incipient degeneration. All stages of transition up to the complete degeneration of the cell can be observed. Shortly before puberty, isolated spermatogonia can be observed in the seminiferous tubule only in rare cases. The spermatogonia from boys aged between six years and puberty generally showed strong signs of degeneration.

According to Hedinger (1976), the deficiency of spermatogonia is, in some cases at least, due not to any secondary disturbance but to hereditary factors. Scorrer and Farrington (1971) see in most cases of cryptorchidism a disturbance of the germ cells resulting from hereditary factors. In our thirteen biopsies from one-year-old boys, there were no visible qualitative differences in appearance between cryptorchid and normal testicles. This would contradict the theory that most cases of cryptorchidism are the result of congenital disturbances of the spermatogonia. It is well known that the number and mitosis rate of spermatogonia is directly related to the gonadotropin level (Bergada and Mancini, 1973; Forest et al., 1973; Forest et al., 1974; Städtler, 1973). The reduced number of spermatogonia already in the first year (Scorrer and Farrington, 1971; Städtler and Hartmann, 1972; Hedinger, 1976) would appear to support the theory that cryptorchidism is caused by gonadotropin deficiency resulting from a disturbance of the hypothalamo-hypophyso-gonadal axis.

2. Sertoli Cells

In cryptorchid testicles too, the cells most frequently encountered in the seminiferous tubule are the Sertoli cells. On the basis of their morphology, two types can be distinguished before puberty, namely the Sa-type and the Sb-type. The fetal Sertoli cells, which in normal testicles persisted until the third month post natum, were already almost non-existent in our material as early as two weeks after birth, the Sa-type Sertoli cells occupying the greater part of the seminiferous tubule. The Sa-type, with its large, round nucleus and only slightly differentiated cytoplasm, shows hardly any qualitative differences from the Sa-type cell of normal testicles, until puberty. The Sb-type, on the other hand, is much rarer and the isolated cells seem to have less smooth endoplasmic reticulum and lipoid droplets than those in normal testicles from the same age-group.

Morphometric assessment of the Sertoli cells up to the age of six revealed no significant difference in number or in individual cell volume. The number of Sertoli cells per cubic centimeter of testicular tissue in normal control testicles was 1.441×10^6, while in cryptorchid testicles the figure was 1.617×10^6. The single cell volume in normal testicles is 491 cu and in undescended testicles 429 cu. The differences in both the number of Sertoli cells and in single cell volume are not significant (Table 2).

In puberty, the transformation of the Sertoli cells to the Sc-type is only partially completed. In most of the biopsies examined, from boys either in puberty or immediately afterwards, the Sertoli cells remained stationary at the Sa-stage. In particular, their nucleus retains its round form and few uncharacteristic cell organelles can be found in the entire cytoplasm. No crystalloid of Charcot-Böttcher is present and the lamellar body is only incompletely developed. The number of lipoid droplets is extremely reduced.

3. Peritubular Connective Tissue

De la Balze et al. (1960) described the changes in the tunica propria of cryptorchid testicles, with particular reference to the fibrotic changes to be found in puberty. The thicker the tubule wall, the more marked were the changes in the tubule, with a resulting decrease in spermatogonia and changes in the Sertoli cells. Sohval (1954) and Mancini et al. (1965), working with the light microscope, noticed changes in the tunica propria in cryptorchid testicles only in puberty. Georgiev and Markov (1970), on the other hand, noticed fibrosis of the tunica propria as early as the age of six, while in eight-year-olds they observed atrophic changes in the germinal epithelium. Only after puberty does the fibrositic tunica propria become hyalinised (Leeson, 1966).

Leeson (1966) carried out electron microscopic studies on cryptorchid individuals between the ages of four and thirty-eight years. A total of twenty-two patients were examined. In none of the children in the prebuberal period was he able to establish any pathological changes at an ultrastructural level. From the age of ten onwards, in cryptorchid individuals, a progressive fibrosis of the peritubular connective tissue sets in, accompanied by delayed maturation of the seminiferous tubule. This fibrosis, which becomes more marked in the third group (fourteen- to thirty-eight-year-olds) is, in his opinion, of particular significance.

The 154 biopsies examined in this study, taken from boys up to sixteen years of age, only partly confirm Leeson's findings. In the first year, the peritubular connective tissue in cryptorchid testicles, as in the normal control testicles, consists of a basement membrane, a collagen fibre zone and a cellular zone. The basement membrane consists of two layers, a lighter, inner layer and a broad, homogeneous, electron-darker outer layer (Figs. 34, 35). The narrow zone of collagen fibres is the same width as that of normal control testicles, with no increased collagenisation. The third, cellular zone, which bounds the tubule towards the interstitium, is composed of two to three rows of fibroblasts. These fibroblasts are spindle-shaped and arranged concentrically around the tubule, with an oval nucleus surrounded by slightly differentiated cytoplasm. There are no signs of degeneration in the cellular zone.

The first changes in the peritubular connective tissue take place in the second year, when the collagen fibre layer shows an increase in collagen fibres, although the layer itself is not yet wider than in normal control testicles of the same age.

In three-year-olds, the changes in the peritubular connective tissue are still more marked. The collagen fibre layer in three-year-old cryptorchid testicles is broader and more collagenised than in controls of the same age (Figs. 36, 37). The fibres lie close together and their diameter appears greater than that of normal testicles. Collagenisation and widening of the tunica propria increase with age in cryptorchid testicles. The biggest differences in the width of the peritubular connective tissue (i. e. the collagen fibre layer) between cryptorchid and normal testicles, are found in seven-year-olds. In fact, the tunica propria continues to become wider with age in both cases, but the dif-

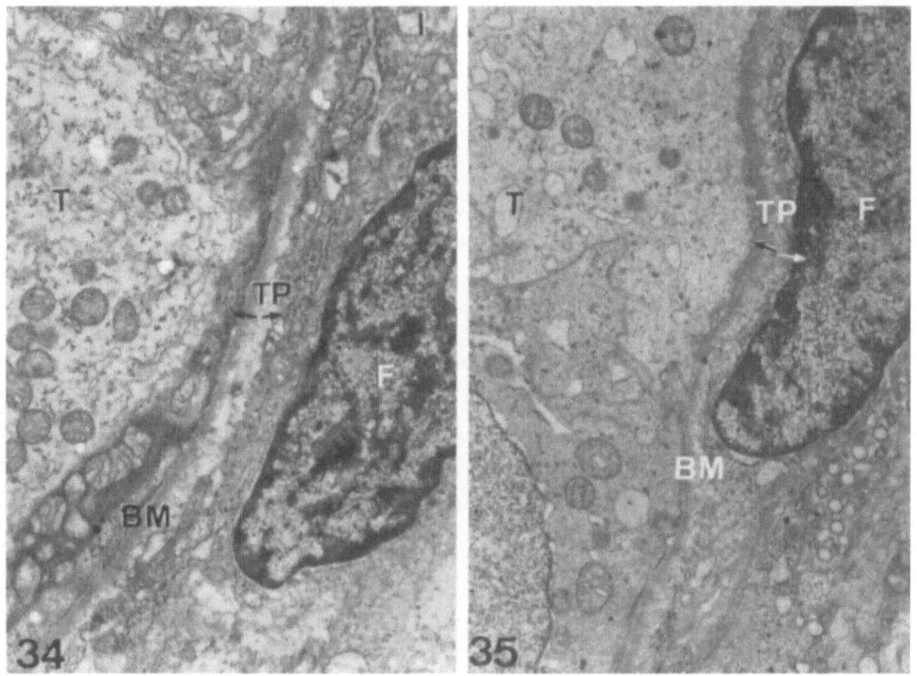

Fig. 34. Normal testicle of a 4-month-old boy. *I* interstitium, *TP* collagen fibre layer is narrow and contains few collagen fibres. *BM* basement membrane, *F* Fibroblast, *T* tubulus. X 9,000

Fig. 35. Cryptorchid testicles of a 3 ½ month-old boy. The collagen fibre layer *TP* is the same thickness as that of a normal boy of the same age. *BM* basement membrane, *F* fibroblast, *T* tubulus. X 9,000

ference between them does not noticeably increase. The density of the collagen fibres and the fibrosis throughout the interstitium show a marked increase between the ages of six and twelve, at which age the whole interstitium appears intensively collagenised, with large accumulations of dense collagen fibres lying between isolated Leydig cells. The basement membrane shows no differences as compared with the normal testicle until puberty, when the normal lamellarisation of the basement membrane does not occur. The otherwise well developed knobs on the basement membrane appear only in rudimentary form and in several tubules the basement membrane exhibits marked irregularities and invaginations.

In fourteen-year-old cryptorchid testicles, the collagen fibre layer has become wider and the collagen fibres are not arranged in orderly fashion. The six to seven cellular layers around the seminiferous tubule are reduced to two to three. The ring of cells, which should entirely surround the tubulus, is imperfectly formed, the cells being grouped together in clumps, their processes only partially developed and the transformation from fibroblasts to myofibroblasts incomplete. The faulty development of the myofibroblasts and the marked fibrosis of the peritubular connective tissue not only make the tubule incapable of contraction: it may also be assumed that the large mucopolysaccharide deposits, induced by the fibrosis, hinder the exchange of substances between the seminiferous tubule and interstitium.

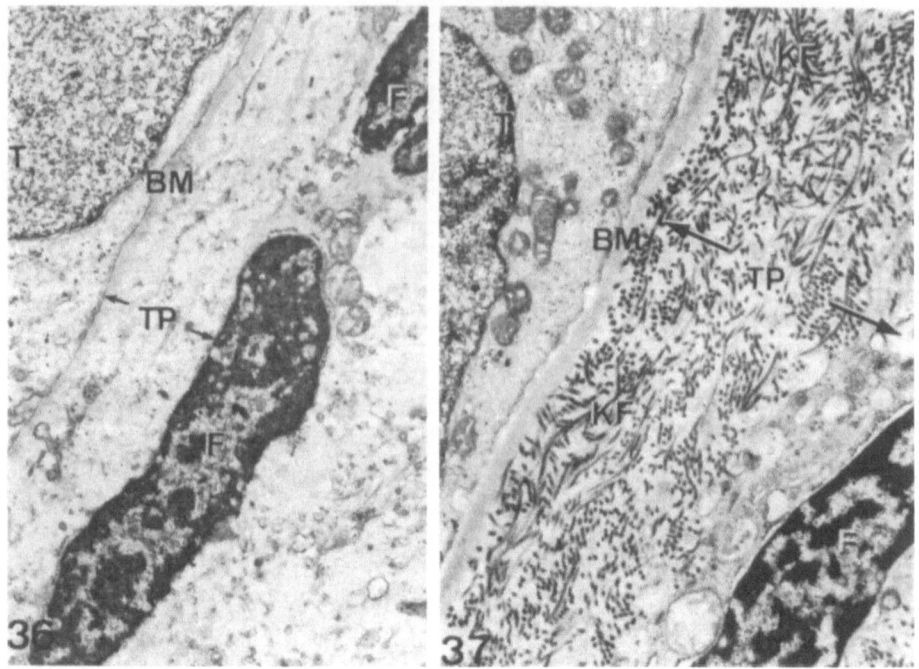

Fig. 36. Normal testicle of a 3-year-old boy. The collagen fibre layer *TP* is narrow. X 9,000. Description, see Fig. 27

Fig. 37. Cryptorchid testicle of a 3-year-old boy. Marked collagenisation and broadening, particularly in the collagen fibre layer. X 9,000. Description, see Fig. 27

The pathological mechanism responsible for the thickening of the wall of the seminiferous tubule in cryptorchidism is unclear. On the one side, gonadotropin deficiency is suggested (Altorfer and Hedinger, 1977), while Usadel et al. (1973) found unmyelinated nerve-fibres in the walls of the testicular tubules, which react positively to cholesterase. It is possible that this is a very active process of penetration. The function of these nerve-fibres, which are observed only in cryptorchids, never in normal testicles, remains unclear (Usadel et al., 1973). Possibly these unmyelinated nerve-fibres are involved in some way in the thickening of the wall of the seminiferous tubule in cryptorchidism.

4. Leydig Cells

Hayashi and Harrison (1971) reported that in cryptorchid testicles no Leydig cells were visible immediately after birth, in contrast to normal control testicles, where Leydig cells could be demonstrated with histological techniques throughout the first year. Many authors were able to establish changes in the Leydig cells only in puberty (Sniffen, 1950; Sohval, 1954; de la Balze et al., 1960; Georgiev and Markov, 1970).

Endocrinological studies showed a reduced production of androgen in cryptorchid testicles in humans (Llaurado and Dominguez, 1941; Engenberg, 1969; Raboch and Starka, 1972; Hooker, 1970).

At the ultrastructural level, Numanoglu et al. (1969) described a progressive degeneration in cryptorchid testicles, beginning from the age of five, and a decrease in the number of Leydig cells compared to normal testicles.

Using the electron microscope, it was possible to confirm, to a certain extent, the observations of Hayashi and Harrison (1971). Unlike them, however, we did find Leydig cells in cryptorchid testicles in one-year-olds, lying singly in the interstitium. Their cytoplasm is much narrower than in normal control Leydig cells from the same age-group. The nucleus is oval, with a prominent nucleolus and peripherally situated nuclear chromatin. Occasionally, empty vacuoles are found in the interior of the nucleus (Fig. 30). When the Leydig cells are markedly atrophic or degenerating, the nucleus is no longer oval, but irregular in shape, with numerous invaginations (Fig. 30). The cytoplasm of these atrophic Leydig cells generally possesses few uncharacteristic cell organelles and fewer mitochondria and lipoid droplets are seen. The smooth endoplasmic reticulum, a typical feature of a fully developed Leydig cell, is very scarce here, vacuoles being more commonly encountered in the cytoplasm of these cells. The degenerating Leydig cells often have only cytolysosomes in their cytoplasm. The atrophic Leydig cells in the cryptorchid testicle described above are strongly reminiscent of the picture of simple atrophy, as described by Hatakeyama (1965). From the first year until puberty, Leydig cells are very seldom found in the interstitium. In puberty, a retarded development of the Leydig cells takes place.

5. Estrogen Induced Cryptorchidism in Mice

a) Newborn Mice

Ultrastructure of Leydig Cells: Raynaud (1942) and Jean (1973) succeeded, after a single dose of estrogen in gravid mice, in producing unilateral or bilateral cryptorchidism in all the male issue. This experiment was repeated and extended to include particularly light and electron microscopic studies of the Leydig cells of these artificially cryptorchid mice. The testosterone content of the newborn and adult mouse testicles was determined and compared with that of normal control testicles of the same age.

On semi-thin sections the Leydig cells from the control mice are easily recognisable, generally lying in groups of three to nine cells in the interstitium. Their cytoplasm is polygonal in appearance. The round nucleus is excentrically situated, with two to three nucleoli. A typical feature of the nucleus of the Leydig cell is the peripherally situated chromatin. In the cytoplasm of one-day-old control mice, numerous lipoid vacuoles are seen, clustered together in groups. The mouse Leydig cells are notably smaller than those of humans.

On the first day post natum, the estrogen-treated mice show small, isolated Leydig cells in the interstitium, lying no longer in groups, but singly. They have a narrow cytoplasm, in which lipoid droplets are seldom present. Besides these atrophic Leydig cells, the interstitium also contains degenerated Leydig cells containing vacuoles. The entire cytoplasm is, to a large extent, filled with large, empty vacuoles. These strongly degenerate Leydig cells were observed only in estrogen-treated cryptorchid mice.

The mice treated i. u. with HCG and estrogen simultaneously have Leydig cells which tend to remain together in groups, unlike those of cryptorchid mice treated with estrogen only. One has the impression that, compared with normal control testicles, the Leydig cells of HCG and estrogen treated mice have noticeably fewer lipoid droplets.

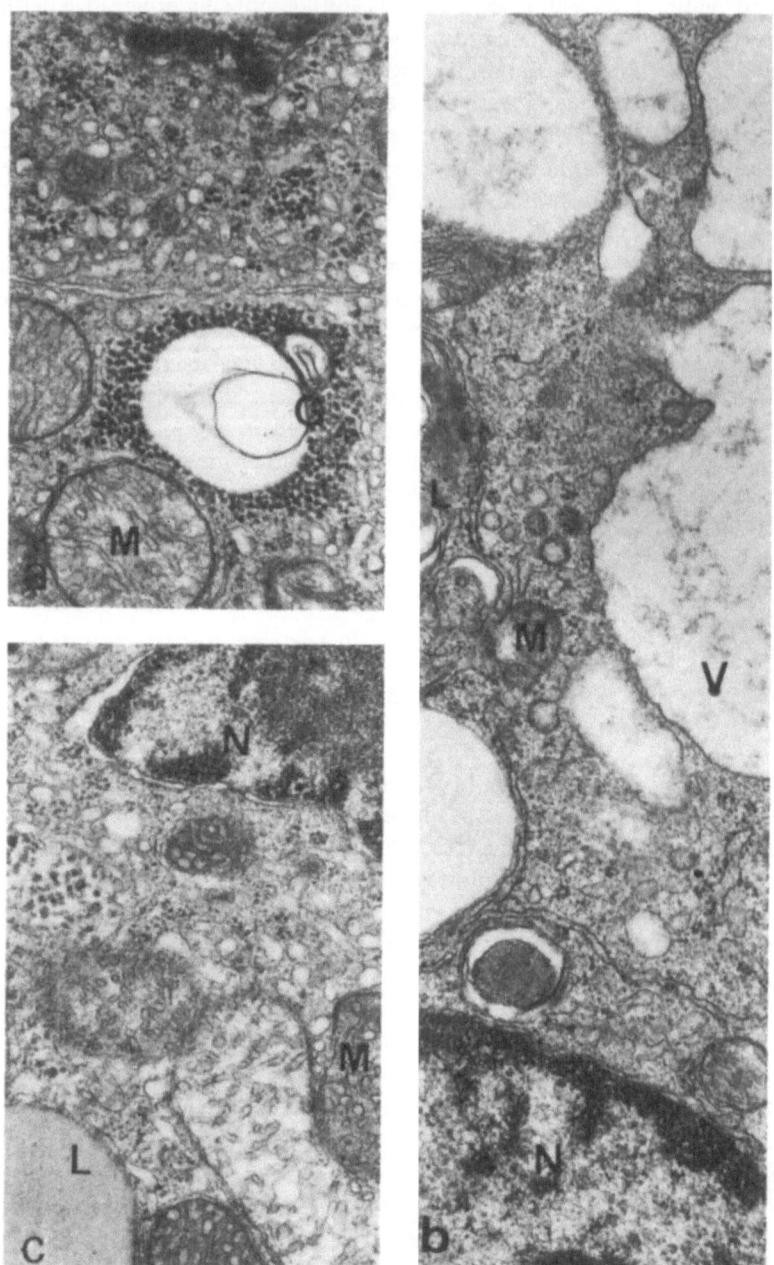

Fig. 38. (a) Electron-microscopic view of the Leydig cells treated with estrogen and HCG during pregnancy. There is an abundance of smooth endoplasmic reticulum *M* mitochondria, *N* nucleus, *G* glycogen. X 31,000. See Fig. 41. (b) Ultrastructure of estrogen-treated Leydig cells. *V* vacuoles, *M* small mitochondria, *N* nucleus, *L* lipoid droplets. X 31,000. See Fig. 40. (c) Cytoplasm of control Leydig cells. *L* lipoid droplets smooth endoplasmic reticulum *M* mitochondria, *N* nucleus. X 31,000. See also Fig. 39

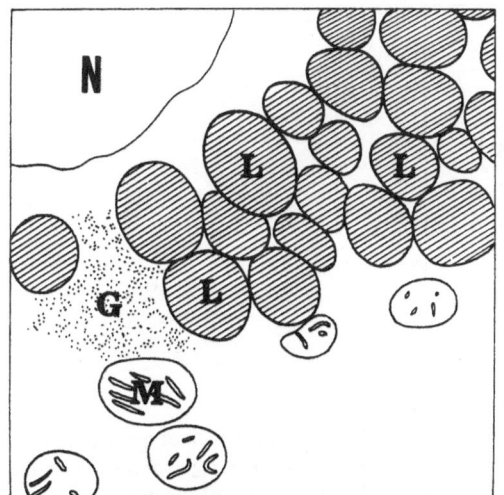

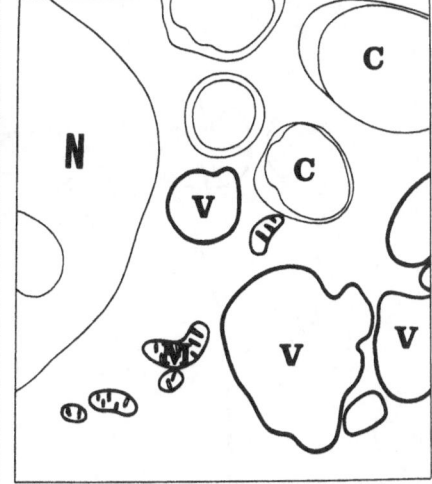

Fig. 39. Normal Leydig cell of a newborn mouse

Fig. 40. Leydig cell of newborn mouse treated with estradiol on the fourteenth day of pregnancy

The ultrastructure of the Leydig cells in the control mice on the first day post natum corresponds largely to the ultrastructure of these cells as described in the literature (Aoki, 1968; Russo, 1971; Russo and de Rosas, 1971; Blackburn et al., 1973). They have an oval nucleus, with heterogenous chromatin situated mainly on the periphery, and loose reticular nucleoli. The most noticeable feature of the polygonal cytoplasm is the abundance of glycogen and ribosomes. The smooth endoplasmic reticulum is moderately well developed: rough endoplasmic reticulum is seldom encountered on the cell periphery (Fig. 38c). The lipoid droplets have a spherical form and lie mostly in groups, rarely singly, and are not in contact with the mitochondria. These latter are numerous, mostly round and of the tubulus type. Intramitochondrial granules are rarely found. The poorly developed Golgi apparatus is found in the vicinity of the nucleus and also on the cell periphery (Figs. 38c, 39).

The Leydig cells of cryptorchid mice whose mothers were treated with estrogen differ markedly from those of the control mice and those of the mice whose mothers received HCG and estrogen simultaneously (Fig. 38b). The cytoplasm is, on the whole, smaller, with a nucleus which is elliptical or fusiform rather than oval and exhibiting invaginations. The chromatin is distributed in groups, still lying mainly around the periphery. One to two loose reticular nucleoli lie on the periphery and form connections with the nuclear membrane. The typical feature of the Leydig cell – its smooth endoplasmic reticulum – is seldom seen here. Glycogen particles are not so abundant as in normal animals on the first day post natum. The mitochondria appear smaller, mostly round, although some are longish in shape. The Golgi apparatus is small and seldom visible in the cytoplasm. The lipoid droplets are reduced in number and have no contact with the mitochondria. In advanced atrophy of the Leydig cells, the endoplasmic reticulum is replaced by vacuoles (Fig. 38b). The mitochondria shrink and cytolysosomes appear in the dark cytoplasm (Fig. 40).

The mice treated with HCG and estrogen have Leydig cells which show signs of stimulation in the cytoplasm (Fig. 38c). The mitochondria are more numerous, larger

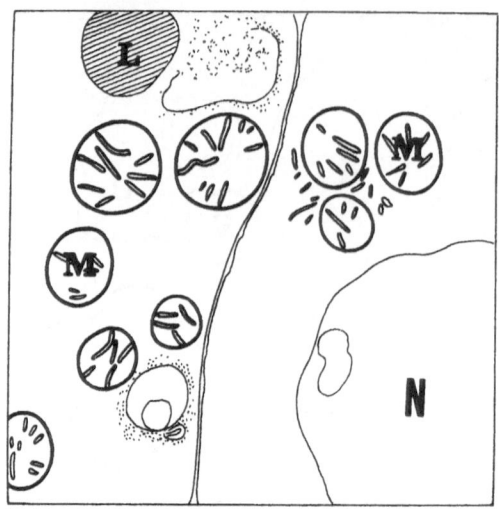

Fig. 41. Leydig cell of a newborn mouse
treated with estrogen and HCG on the
fourteenth day of pregnancy

in appearance than those of the control animals and have a tendency to form groups.
Intramitochondrial granules are frequently visible. Lipoid droplets, in contact with
the mitochondria are less common. The smooth endoplasmic reticulum is more highly
developed, lying in vesicular and tubular form in the cytoplasm. The rough endoplasmic reticulum, too, is more frequently encountered and the glycogen forms larger
agglomerations (Fig. 41). The Golgi apparatus is also larger and more differentiate.

b) Adult Mice

Ultrastructure of the Leydig Cells: Normal Leydig Cells: The epitheloid cells of adult
control mice are polygonal in shape, with an excentrically located nucleus, which
varies in shape, but is often ovoid. The Leydig cell surface contains irregular microvilli, intercellular canaliculi and special septate type junctions (Aoki, 1968). The
smooth endoplasmic reticulum is the most striking feature of the Leydig cells. It occupies the greater part of the cytoplasm and is mostly tubular in form.

In the cytoplasm there are usually areas where the smooth endoplasmic reticulum is
organised into closely packed, fenestrated cisternae. Occasionally the lamellae of the
smooth endoplasmic reticulum form lamellae in the paracrystalline areas (Fig. 42). In
the vicinity of the concentric rows of smooth endoplasmic reticulum, there are many
granules, 2500 Å diameter, round or oval in shape and bounded by a membrane, with
homogeneously dispersed electron-dense contents.

An accumulation of these granules is also seen around the intercellular canalicules.
The mitochondria are about 1 μ in diameter, generally rod-shaped. They are of the
crista type, but tubuli are also observed, some of them with intramitochondrial granules. The microvilli and mitochondria are in close contact. The numerous, membrane-
bounded lipoid droplets are arranged in groups, with smooth endoplasmic reticulum
arranged in concentric circles around them. Pigment granules, presumably lipofuscin,
are also occasionally seen in the cytoplasm. Adjacent to the nucleus is a prominent
centrosomal region. The perinuclear region is relatively free of smooth endoplasmic
reticulum and contains ribosomes and glycogen granules. No crystalloids are observed
in mouse Leydig cells.

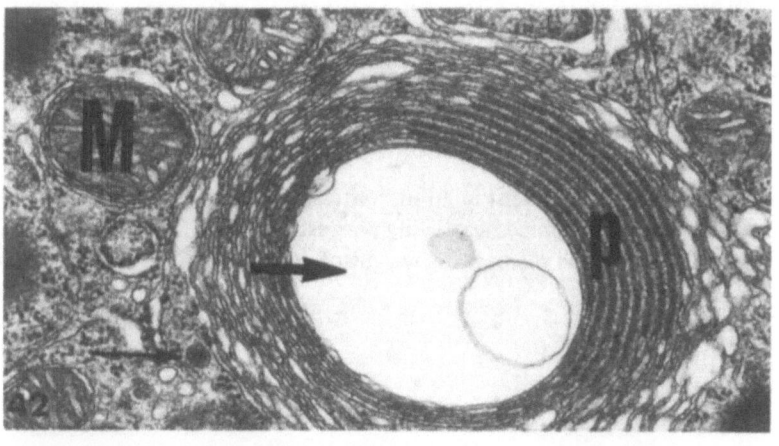

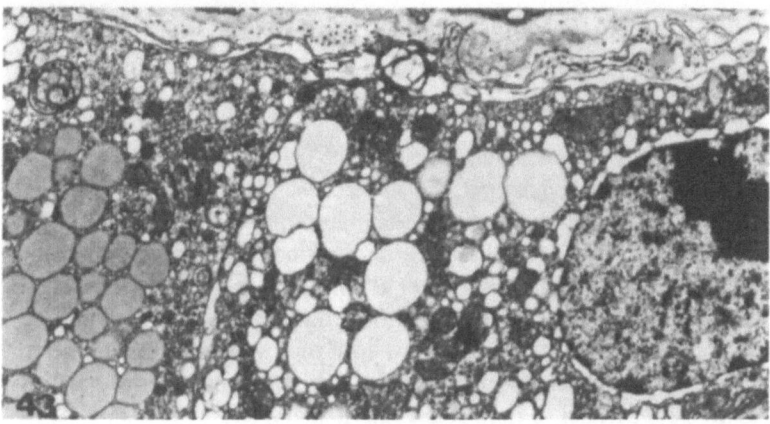

Fig. 42. Normal Leydig cell cytoplasma of adult mice. A smooth endoplasmic reticulum lies in tubular and vacuolar form in cytoplasm. *P* paracrystaline area, † Microbodies, *M* Mitochondria, † Lipoid droplets. X 20,000

Fig. 43. Leydig cells of estrogen treated adult mice. The smooth endoplasmic reticulum is enlarged and builds vacuoles. The lipoid droplets content seems to be increased. The prominent nucleolus is situated peripherally in an excentrically located nucleus. X 10,200

Leydig Cells of E_2B Adult Mice: The Leydig cells of E_2 B mice generally appear different from those of control animals. The epitheloid cells are generally smaller, with a characteristic cytoplasm. The smooth endoplasmic reticulum is replaced by many vacuoles of various sizes, scattered throughout the cytoplasm. The lipoid content of these cells seems to be increased (Fig. 43). A prominent nucleolus is often noted in the excentrically located nucleus, which is more differentiated than that of control Leydig cells. The granules, 2500 Å in diameter, frequently observed in control Leydig cells, are here almost completely absent. The glycogen and ribosome content is greatly reduced. The mitochondria are of the crista type and generally smaller. Many pigment inclusions, presumably lipofuscin, are accumulated in the cytoplasm. The cell surface is not differentiated and there are only rudimentary microvilli. These cells often degenerate, so that we observed in the interstitium many nuclei without cytoplasm and large numbers of plasma bodies.

c) Testosterone Content

The testosterone content from cryptorchid mouse testis was 49 pg/testis, whereas the newborn controls had a testosterone content of 106 pg/testis (P < 0.0001). The weight difference of newborns in E_2B mice (1.29 g) and control mice (1.22 g) was not significant.

The total testicular testosterone content in adult controls was 72.5 ng per testis and in E_2B, i. u. estrogen treated adults mice, was 19 ng per testis (P < 0.001). There is no significant difference between the mean testis weight of control mice (129 g) and that of E_2B mice (126 g).

Discussion

The most widely held theory of. testicular development from birth to puberty was that of Charny et al. (1952), who distinguished three phases in the development of the testicle: the static phase, the growth phase and the maturation phase. Städtler (1975) rejected the theory of development in phases, on the grounds that the seminiferous tubule showed a continuous increase in diameter and in the number of spermatogonia up to the eleventh year. His results have been confirmed in this study. The diameter of the seminiferous tubule increases in linear fashion until the twelfth year, when a rapid increase takes place, following the appearance of the lumen. The number of spermatogonia shows an increase parallel to that of the tubule diameter.

Vilar (1970) described the Sertoli cells as immature supporting cells, whose number increased steadily until puberty, atwhich point, and not before, cell division ceased. Our own studies show that the number of Sertoli cells per tubular cross-section decreases from birth to puberty and no division of the Sertoli cells between birth and puberty was ever observed in any of the biopsies examined.

The Sertoli cells not only have a supporting function: they also play nutritive, phagocytizing and hormone-producing roles. The fetal Sertoli cells are specialised cells, with abundant smooth and rough endoplasmic reticulum, lipoid droplets and lamellar bodies. As such, they are either under the direct influence of gonadotropin or are indirectly stimulated by the androgens. After birth, when gonadotropin stimulation ceases and the androgen level drops, the fetal Sertoli cells change into Sa-type cells, remaining as undifferentiated cells until puberty.

In addition to the Sa-type, Sb-type cells are found throughout childhood. These latter are ultrastructurally more differentiate than the Sa cells and play a phagocytizing as well as a supporting role. With the rise in gonadotropin level in puberty, the Sa and Sb cells change into Sc cells. Except for the fact that the Sc cell is five to six times larger than the fetal Sertoli cell, the two have many similarities. Here again, we encounter lamellar bodies, a well developed smooth and rough endoplasmic reticulum, an abundance of lipoid droplets and longish mitochondria. The whole cell shows a tendency to polarisation and nuclear invaginations. These common features may be attributed to the direct influence of gonadotropin or to direct androgen stimulation.

The second commonest type of cell in the juvenile seminiferous tubule are the germ cells. Immediately after birth, gonogytes can be seen in the tubule, in addition to the spermatogonia. These gonocytes are structurally similar to the fetal gonocytes, showing also a postnatal tendency to migrate towards the basement membrane. The same, pseudopodia-like movements of the gonocytes were described in the fetus by Gondos and Hobel (1971). Some of the gonocytes change into spermatogonia upon coming in contact with the basement membrane, while the rest degenerate and are, in part, phagocytized by the fetal Sertoli cells (Hadžiselimović, 1976). Roosen-Runge and Leik (1968), with the light microscope, were also able to observe in rats a process of gonocyte degeneration followed by phagocytosis by the Sertoli vells. The spermatogonia count shows a steady increase from birth to the twelfth year. According to Koch et al. (1975), the spermatogonia decrease in the second year and then increase steadily until puberty. These observations do not agree with those of Mancini et al. (1965), who noticed no drop in the spermatogonia count in the second year. One explanation of this discrepancy may be that, while Mancini et al. made a distinction between gonocytes and spermatogonia in the first year, Koch et al. did not do so.

A generally accepted view was that only so-called "prespermatogonia" and gonocytes are present in the child seminiferous tubule (Vilar, 1970), the Ap, Ad and B spermatogonia making their appearance only in puberty. The observations reported in "The Normal Testicle", p. 10 show that Ap and Ad spermatogonia are present in the seminiferous tubule as early as the fifth week post natum. In the fifth year, B spermatogonia and primary spermatocytes make their appearance. Spermatogenesis then remains at the stage of primary spermatocytes until puberty. According to Städtler (1975), a possible explanation of this discrepancy might be that Vilar carried out his studies on pathological material. Niemi and Ikonen's observation (1965) that fetal and transitional spermatogonia are present until the seventh year, was confirmed by electron microscopic studies.

A characteristic feature of adult, mature spermatogonia, Sertoli and Leydig cells is the appearance of crystalloids in their cytoplasm. In the case of the spermatogonia, the crystalloid is present as early as the fifth week post natum, while. Sertoli and Leydig cells acquire their crystalloid only at puberty. From the point of view of structure, these crystalloids are of three different types. In children, they are ultrastructurally similar to those found in the adult spermatogonia, Sertoli and Leydig cells.

The peritubular connective tissue, from immediately after birth till puberty, consists of three zones: 1. basement membrane, 2. collagen fibre zone, 3. cellular zone.

This last is composed of fibroblasts, two different types being distinguishable: Type I, a younger fibroblast and Type II, an older fibroblast. Type II is identical with Fawcett and Burgos's (1960) precursor Leydig cell (Hadžiselimović and Seguchi, 1975). In puberty, under the influence of gonadotropin, the Type I fibroblasts change into myofibroblasts.

The fourth element of the testicle, the Leydig cells, are situated around the vesicles, between the tubuli. According to Vilar (1970), the Leydig cells degenerate irreversibly in the first three months post natum, no more Leydig cells being found in the interstitium in the third year. Hatakeyama (1965), on the other hand, makes a clear distinction between simple atrophy of the Leydig cells and actual degeneration. Leydig cells in a state of simple atrophy can be found throughout the whole period of childhood, and it is from them that the adult Leydig cells develop in puberty. These results were partly confirmed by Hayashi and Harrison (1971). According to these authors, the appearance of the Leydig cells can be traced in phases from birth to puberty. From birth to the end of the first year, the Leydig cells can be clearly distinguished with the light microscope, disappearing from the interstitium in the second year, to reappear between the age of five and six, after which they remain in the interstitium until puberty. Ultrastructural studies have confirmed these light-microscopic observations. The Leydig cells are very well developed in the first year, after which they enter an inactive phase, which lasts till the beginning of puberty. The inactive Leydig cells or those in the stage of simple atrophy are found increasingly in the interstitium between the age of five and eight. Parallel to this process, B spermatogonia and primary spermatocytes make their appearance in the seminiferous tubule. This parallel is morphological in nature and cannot be demonstrated endocrinologically. No rise in gonadotropins or androgens can be detected in this period. Between the ages of two and three and eight and nine, the juvenile Leydig cells are very seldom seen in the interstitium, precursor Leydig cells being much more common in this period. With the onset of puberty, the Leydig cells become larger and the amount of smooth endoplasmic reticulum shows a marked increase. Until the age of thirteen, no Reinecke's crystalloid was found in the Leydig cells studied.

At the ultrastructural level, it is possible to divide the changes taking place in cryptorchidism into primary and secondary. The secondary changes, which are first seen in the peritubular connective tissue, most certainly play an important role in connection with the subsequent fertility of cryptorchid boys. As early as the second year, collagenisation sets in in the collagen fibre layer and by the age of three this zone shows a marked increase in size compared with normal control testicles from the same age-group. This collagenisation and expansion of the peritubular connective tissue becomes more marked with increasing age.

The literature offers a somewhat confusing survey of the secondary changes in the spermatogonia. Farrington (1969) is of the opinion that cryptorchid spermatogonia are, in fact, congenitally damaged, while Hedinger (1971), Hoesli (1971), Bodenski and Regele (1973) and Jendricke et al. (1973) recorded an almost normal spermatogonia count in one-year old cryptorchid boys. No studies have yet been made to determine the nucleus-plasma ratio and the appearance of atrophy in the spermatogonia as age increases. Our own investigations on spermatogonia from biopsies of one- and three-year-old cryptorchid boys showed a significant decrease in the cytoplasm in the third year, while the nucleus remains almost the same size. These histometric studies show clearly that the secondary changes, in the form of atrophy of the spermatogonia, are already very marked by the age of three. While a significant decrease in the volume density of the spermatogonia takes place up to the sixth year, the single cell volume and the volume density of the Sertoli cells remains unchanged up till this point, although they too show signs of alteration with increasing age. Thus, after puberty, it is common to find only Sa- and Sb-type cells, the transition to Sc-type cells taking place only very seldom. The incomplete transformation of the Sa- or Sb-type Sertoli cells to the Sc-type could be ascribed to gonadotropin deficiency or lack of gonadotropin stimulation.

With regard to the primary changes, which affect mainly the Leydig cells, it is important to note that these atrophic Leydig cells can bee seen immediately after birth in the interstitium of cryptorchids. While Hayashi and Harrison (1971) could find no Leydig cells in one-year-old cryptorchid boys, I was able to locate them in all the biopsies of cryptorchids of this age-group. The atrophic appearance of these Leydig cells compared with the well developed Leydig cells found in normal individuals of the same age supports the theory of deficient stimulation.

Observations made in experimentally cryptorchid mice reveal that animals showing atrophy of the Leydig cells also have a reduced testicular testosterone content, thus indicating that a relationship exists between endocrine activity and the ultrastructural appearance of the Leydig cells. This reduced androgen content was observed on several occasions in cryptorchid humans and also in the pig (Hanes and Hooker, 1939). Sniffen (1950) observed pathological Leydig cells in puberal cryptorchid boys. The ultrastructural changes in the Leydig cells of newborns are identical with those found in newborn mice, where cryptorchidism was induced experimentally. The high temperature, which, according to Sniffen, is the cause of atrophy of the Leydig cells in ectopically situated testes, cannot explain the atrophic changes in newborn children and artificially cryptorchid mice. Neither can the functional and morphological pathology of the Leydig cells of adult mice, whose mother received estradiol on the fourteenth day of gestation, be ascribed to the direct effect of estradiol. Further, the atrophic changes in the Leydig cells of newborn mice can be largely eliminated by simultaneous administration of HCG and estrogen (Hadžiselimović and Herzog, 1975). All these facts indicate that the normal functioning of the Leydig cells is a decisive factor in descensus testiculorum.

Leydig cell function is under the control of pituitary LH. It is obvious, therefore, that impaired LH stimulation could be a pathogenetic factor in the morphological and functional changes in inactive Leydig cells. It must thus be assumed that intrauterine injection of estradiol affects the gonadal system at the hypothalamic or pituitary level. This leads to impaired androgen production in the fetal testis, which, in turn, is responsible for maldescensus. The findings in adult estrogenized mice demonstrate that the suppression of the hypothalamo-pituitary-gonadal system is not transient, but permanent. The findings of Job et al. (1974) that cryptorchid children show an impaired LH secretion in response to LH-RH and the clinical observations of Praeder et al. (1976) are readily explained by the present experimental study.

In the sixty-six cases of cryptorchidism operated by myself in 1975, a proper mechanical barrier was found in only two cases, where the route to the scrotum had to be remade. It is possible that in these two boys the cause of cryptorchidism was of a purely mechanical nature.

Therapy

The main problem in the therapy of cryptorchidism is how to preserve fertility. In this connection, attention today is undoubtedly focussed on the question of early treatment. The fact that this is regarded as imperative is based on the following considerations: The secondary changes in cryptorchidism are clearly visible at the ultrastructural level as early as the second year. Collagenisation and widening of the tunica propria, which assume greater proportions with increasing age, are accompanied by the disappearance of spermatogonia from the seminiferous tubule. The significant fall in the spermatogonia cell-count from the second year onwards has, in the meantime, been confirmed by several authors (Hedinger, 1971; Hoesli, 1971; Hecker et al., 1972; Städtler and Hartmann, 1972; Jendricke, 1976). At the same time, the tubules decrease in diameter and take no part in puberal development (Hecker 1977). The atrophic changes are very marked in the third year, as histometric examinations carried out in our clinic confirm (Lüdin, 1976). The decisive question, namely whether the spermatogonia recover and increase in number after operation was put forward by Kleinteich and Schickedanz (1976), who carried out histological studies on 157 testicles from patients who had undergone operation for unilateral or bilateral cryptorchidism two to ten years previously. The mean spermatogonia content per cross-section of one seminiferous tubule in each testicle was determined from fifty circular tubulus cross-sections. The value for the treated testicles was then compared with that obtained from untreated cryptorchid testicles in the same age-group and was found to be appreciably higher in the operated testicles than in comparable untreated testicles.

The earlier the operation or treatment is carried out, the better the results with regard to fertility and the greater the increase in the spermatogonia count. It is even possible to obtain values comparable to those for normal testicles (Ludwig and Potempa, 1975).

The atrophy of the Leydig cells, which is already visible in cryptorchid testicles at birth, together with the experimental studies described in Chapter 5 of this work, is evidence of the existence of an endocrinological disturbance in cryptorchidism. Knorr (1969) rightly described cryptorchidism as the most common disorder of the endocrine glands. The successful preservation of fertility achieved with HCG therapy is further evidence of the involvement of endocrine factors in maldescensus. Hormone therapy of cryptorchidism is thus to be preferred to operative treatment. The following recommendations for the therapy of cryptorchidism are the proven results of this monograph:

Programme of Therapy

1. HCG therapy should be started in infancy, after the age of six months in both unilateral and bilateral cryptorchidism.
   Dosage: 2 x 250 I. U. per week over a period of 5 weeks.

2. Should hormone therapy prove unsuccessful, an operation should be carried out by an experienced pediatric surgeon, at the age of two.

3. In cases where neither testicle can be located by palpation, the plasma LH and FSH, possibly also the LH-RH test and the testosterone content of the plasma must be determined. In addition, in all newborns with anorchidism, a sex-chromatin count should be made.

4. In every case of maldescensus of the testicles in newborns, the parents must be made fully aware of the importance of check-ups and, where appropriate, early treatment of the condition.

Summary

1. The seminiferous tubule in children consists of Sertoli cells, gonocytes, spermatogonia and degenerating cells.

2. The Sertoli cells are the commonest cells in the child seminiferous tubule and can be subdivided into fetal, Sa-type, Sb-type and Sc-type cells.

3. The gonocytes are visible until the third month, when they give rise to fetal spermatogonia. In addition to these fetal spermatogonia, transitional, Ap, Ad and B spermatogonia can all be found in children. The Ap and Ad spermatogonia are present as early as the fifth week post natum, while B spermatogonia and primary spermatocytes do not make their appearance until the age of five. Spermatogenesis remains stationary at this stage until the start of puberty. In contrast to the Sertoli cells, the number of spermatogonia increases in an apparently linear fashion from birth to puberty.

4. The peritubular connective tissue is composed of the basement membrane, a collagen fibre layer and a cellular layer. In puberty, under the influence of gonadotropin, the basement membrane becomes multi-layered and the fibroblasts change into myofibroblasts.

5. Well-developed Leydig cells are present in the interstitium in the first year, after which the juvenile stage sets in, persisting until puberty.

6. The changes occurring in cryptorchidism can be divided into primary and secondary changes.

7. The primary changes affect the Leydig cells, atrophic Leydig cells being visible in cryptorchid testicles as early as one week post natum.

8. Reduced gonadotropin stimulation i. u. is considered one of the main factors in the etiology of cryptorchidism.

9. The secondary changes are visible for the first time in the second year, in the form of collagenisation and widening of the collagen fibre layer of the peritubular connective tissue. These changes are accompanied by atrophy of the spermatogonia. The Sertoli cells show no changes in cell volume or number until the age of six. After puberty most Sertoli cells are still at the Sa-stage.

10. In experimentally induced cryptorchidism in mice, the young animals showed atrophically altered Leydig cells identical to those found in newborn boys. The testosterone content of such testes is significantly lower than that of control testes from the same age-group. Even in adult mice treated i. u. with estradiol, the testosterone content is significantly diminished. This experiment supports the hypothesis that cryptorchidism results from an insufficiency in the hypothalamo-hypophysogonadal system.

11. Early treatment of cryptorchidism is advocated: gonadotropin therapy in the first year and operative in the second.

Acknowledgements

I would like to take this opportuntity to thank the following for their help in the preparation of this monograph:

My teacher, *Prof. R. Nicole,* Professor emeritus Pediatric Surgery, who stimulated my interest in cryptorchidism and inspired me to write this monograph.

Prof. B. Herzog, Professor Pediatric Surgery at the University of Basle, and successor of Prof. R. Nicole, for the continued support of this work and for the biopsy material.

Prof. G. Stalder, Director of the Pediatric Clinic, Basle and *Prof. G. Wolf-Heidegger,* Director of the Institute of Anatomy, Basle for his help and support in my work.

Prof. H. U. Zollinger, Director of the Institute of Pathology and *Prof. H. P. Rohr,* Head of the Laboratory for Electron Microscopy at the Institute of Pathology, University of Basle for enabling me to carry out the electron microscopic examinations and morphometric studies reported here.

Prof. J. Girard, Head of the Endocrine Laboratory of the University Childrens' Hospital of Basle for valuable suggestions and friendly support.

Prof. Th. Schiebler, Director of the Institute Anatomy of the University of Würzburg for his advice in the preparation of this monograph.

Miss G. Krey of the Institute Pathology in Basle for her skilled technical assistance in the preparation of the testicular biopsies.

Frau N. Lochbrunner for the English translation of the text.

Mr. R. Muspach, technical drawer at the University of Basle, for the preparation of the tables and for the excellent diagrams which help to make the electron micrographs more understandable to the reader.

Miss D. Kohler and *Miss M. Mangold* for their help in typing this monograph.

Finally, I would like to thank particularly all those who have helped me directly or indirectly in this work.

References

Aoki, A.: Hormone-induced differentiation of agranular endoplasmic reticulum in the interstitial cells of the mouse testis. Protoplasma **66**, 263–267 (1968)

Albert, A., Underdahl, L. O., Greene, L. F., Lorenz, N.: Male hypogonadism I. The normal testis. Mayo Clin. Proc. **28**, 409–422 (1953)

Altdorfer, J., Hedinger, Chr.: Diagnostische Bedeutung elektronenmikroskopischer Untersuchungen der Tunica propria der Samenkanälchen. Schweiz. Pathol. Tag. 1974 In Press.

Baille, H.: Ultrastructural differentiation of the basement membrane of the mouse seminiferous tubule. Quart. J. Micro. Sci. **105**, 203–207 (1964)

Balze, F. A. de la, Mancini, R. E., Arrillaga, F., Andrada, J. A., Vilar, O., Gurtmann, A. I., Davidson, O. W.: Histologic study upon the undescended human testis during puberty. J. clin. Endocr. **20**, 286–297 (1960)

Bawa, S. R.: Fine Structure of the Sertoli cell of the Human Testis. J. Ultrastruct. Res. **9**, 459–474 (1963)

Bayle, H.: De traitment chirurgical des cryptorchidies. In: La founction endocrine du testicle. Paris: Masson 1957

Bergada, C., Mancini, R. E.: Effect of gonadotropins on the induration of spermatogenesis in human prepuberal testis. J. clin. Endocr. **37**, 935–943 (1973)

Blackburn, W. R., Kyung, W. C., Bullock, L., Bardin C. W.: Testicular feminization in the mouse. Studies of Leydig cell structure and function. Biol. Reprod. **9**, 9–23 (1973)

Bodensky, G., Regele, H.: Histologische Untersuchungen beim Leistenhoden. Mschr. Kinderheilk. **121**, 611–613 (1973)

Bouin, P., Ancel P.: Recherches sur le cellules interstitielles du testicles chez les mammiferes. Arch. Zool. exp. Gen. **1**, 437–523 (1903)

Burgos, M. H.: Fine structure of basement membrane of the human seminiferous tubulus. Anat. Rec. **136**, 312 (1960)

Burgos, M. H., Fawcett, D. W.: Studies on the fine structure of the mammalian testis. J. biophys. biochem. Cytol. **1**, 287–299 (1955)

Canlorbe, P., Toutblanc, J. E., Job, J. C., Scholler, R., Roger, M., Castanier, M.: Fonction endocrine du testicule dans 125 cas de cryptorchidies. Ann. Endocr. (Paris) **35**, 177–181 (1974)

Carver, J. H.: Bilateral orchidopexie and fertility. Proc. roy. Soc. Med. **51**, 328–330 (1958)

Charny, Ch. W., Conston, A. S., Meranze, D. R.: Development of the testis. Fertil. and Steril. **3**, 461–479 (1952)

Clermont, Y.: Contractile elements of basement membrane of the human seminiferous tubulus. Anat. Rec. **136**, 312–316 (1960)

Court, M., Hochereau-de Reviers, M-T., Ortavant R.: Spermatogenesis. In: Testis I. (eds. Johnson, A. D., Gomes, W. R., Vandemark, N. L.) New York, London: Academic Press p. 339–432, 1970

Diamentopoulus, S.: Ueber die Hypoplasie der Hoden in der Entwicklungsperiode. Z. Ges. Anat. **8**, 117–126 (1921)

Doepfmer, R., Nienaber, W.: Die einseitige Hodendystopie. Münch. med. Wschr. **106**, 2096–2101 (1964)

Dorrington, J. H., Fritz, B., Armstrong, D. T.: Testicular estrogens: synthesis by isolated Sertoli cells and regulation by follicles-stimulating hormone. In: Regulatory mechanisms of male reproductive physiology. (eds. Spielman, C. H., Lobl, T. J., and Kirton, K. T.) p. 63–70 Excerpte medice, Amsterdam, Oxford, New York: American Elsevier 1976

Ebner, H. von: Untersuchung über den Bau der Samenkanälchen und die Entwicklung der Spermatozoiden bei den Säugetieren und beim Menschen. Rollet's Untersuch. Inst. Physiol. **200**, 1–43 (1871)

Eik-Nes, K. B.: Secretion of testosterone by the atopic and cryptorchid testes in the same dog. Can. J. Physiol. Pharmacol. **44**, 629–633 (1966)

Engberg, H.: Investigation of the endocrine function of the testicle in cryptorchidism. Proc. roy. soc. Med. **42**, 652–655 (1949)

Farrington, G. H.: Histologic observations in cryptorchidism: The congenital germinalcell dificiency of undescended testis. J. pediat. Surg. **4**, 606–613 (1969)

Fawcett, D. W., Burgos, M. H.: The fine structure of Sertoli cells in human testis. Anat. Rec. **124**, 401–402 (1956)

Fawcett, D. W., Burgos, M. H.: Studies on the fine structure of the mammalian testis. II The human interstitial tissue. Amer. J. Anat. **107**, 245–269 (1960)

Fawcett, D. W., Leak, L. V., Hediger, P. M.: Electron Microscopic Observations on the Structural Components of the Blood-Testis Barrier. J. Reprod. Fertil. Suppl. **10**, 105–122 (1970)

Forest, M. G., Cothiard, A. M., Bertrand, J. A.: Evidence of testicular activity in early infancy. J. clin. Endocr. **37**, 148–151 (1973)

Forest, M. G., Sizonenko, P. C., Cothiard, A. M., Bertrand, J.: Hypophyses-gonadal function in human during the first year of life. J. Clin. Invest. **53**, 819–828 (1974)

Fukuda, T., Hedinger, Chr., Grosscurth, P.: Ultrastructure of developing germ cells in the fetal human testis. Cell Tiss. Res. **161**, 55–70 (1975)

Gardner, P. G., Holyoke, E. A.: Fine structure of the seminiferous tubule of the Swiss mouse. I The Limiting membrane, Sertoli cells, spermatogonia and Spermatocytes. Anat. Rec. **150**, 391–404 (1964)

George, U.: Beobachtungen an Sertoli-Zellen im Hoden des Menschen und der Ratte. Z. mikrosk.-anat. Forsch. **42**, 479–498 (1937)

Georgiev, G., Markow, D.: Histological changes in undescended testis. Scr. sci. med. Verena 8, 109–111 (1970)

Gondos, B., Hobel, C. J.: Ultrastructure of germ cell development in the human fetal testis. Z. Zellforsch. **119**, 1–20 (1971)

Guillon, G., Seguy, E.: Cryptorchidisim and male infertility. Coloque sur les cryptorchidies. Symposion Inter. Fert. Ass. Amsterdam. Das Hormon 4 (1964)

Hadžiselimović, F., Seguchi, H.: Elektronenmikroskopische Untersuchungen an Kinderhoden bei unvollständigem Descensus. Acta anat. Basel **86**, 474–483 (1973)

Hadžiselimović, F., Seguchi, H.: Ultramikroskopische Untersuchungen am Tubulus Seminiferus bei Kindern von der Geburt bis zur Pubertät. II Entwicklung und Morphologie der Sertoli-zellen. Anat. Anz. Erg. **68**, 149–161 (1974)

Hadžiselimović, F., Seguchi, H.: Razvoj testisa u djece. Fol. anat. jug. **1**, 67–74 (1974)

Hadžiselimović, F., Seguchi, H.: An electron microscopical study on the postnatal development of human Leydig cells. 10th Int. Cong. Anat. Tokyo 445 (1975)

Hadžiselimović, F., Herzog, B., Seguchi, H.: Surgical correction of cryptorchidism at 2 years. Electron microscopic and morphometric investigations. J. pediat. Surg. **10**, 19–28 (1975)

Hadžiselimović, F., Herzog, B.: The Meaning of the Leydig cell in Relation to the Etiology of Cryptorchidism. J. pediat. Surg. **11**, 1–8 (1976)

Hadžiselimović, F.: Elektronsko mikroskopska proucavanja promjena na goncitima djece neposredno poslije rodjenja. Fol. anat. jug. **5**, 37–43 (1976)

Hadžiselimović, F.: Elektronenmikroskopische Untersuchungen über die Entwicklung des Tubulus seminiferus unmittelbar nach der Geburt bis zum vollendeten ersten Lebensjahr. Acta anat. Basel **95**, 287–299 (1976)

Hadžiselimović, F., Girard, J.: Pathogenesis of cryptorchidism. Hormone res. 1977 In press

Hamilton, J. B., Leonard, S. L.: The effect of male hormone substance upon the testes and upon spermatogenesis. Anat. Rec. **71**, 105–117 (1938)

Hamilton, J. B.: The effect of male hormone upon the descent of the testes. Anat. Rec. **70**, 533–541 (1939)

Hanes, F. M., Hooker, C. W.: Hormone production in the undescended testis. Proc. Soc. exp. Biol. (N. Y.) **35**, 549 (1937)

Hansen, T. S.: Fertiliteten ved operativt behandlet og ubehandlet kryptorchisme. Med. Diss. Kopenhagen 1945

Hatakeyama, S.: A study on the interstial cells of the human testis, especially on their fine-structural pathology. Acta path. jap. **15**, 155–197 (1965)

Hayashi, H., Harrison, R. G.: The development of the interstitial tissue of the human testis. Fertil. and Steril. **22**, 351–355 (1971)

Hecker, W. Chr.: Neue Gesichtspunkte zum Kryptorchismusproblem. Münch. med. Wschr. **113**, 1125–1128 (1971)

Hecker, W. Chr., Knorr, D., Mengel, W., Moritz, P.: Die Behandlung des Maldescensus testis unter besonderer Berücksichtigung des Operationszeitpunktes. Kolloqium Maldescensus testis. Tübingen 1976. In Press

Hecker, W. Chr.: Hodenhochstand, Fakten zur Frühtherapie und Ueberlegungen zur Aetiologie. Kinderarzt 1, 13–17 (1977)

Hedinger, Chr.: Ueber den Zeitpunkt frühest erkennbarer Hodenveränderungen beim Kryptorchismus des Kleinkindes. Verh. dtsch. Ges. Path. 55, 172–175 (1971)

Hedinger, Chr.: Die Histologie des kryptorchen Hodens. Kolloquium Maldescensus testis. Tübingen 1976. In Press

Hellinga, G.: Fertility after hormonal or surgical treatment from bilateral cryptorchidism. Symposium Int. Fertil. Ass. 25, 4, 1964 Amsterdam. Das Hormon 7 (1964)

Holstein, A. F., Wartenberg, H., Vossmeyer, J.: Zur Cytologie der pränatalen Gonadenentwicklung beim Menschen. Z. Anat. Entwickl. Gesch. 135, 43–66 (1971)

Hooker, W. Ch.: The intertubular tissue of the testis. In: Testis I (eds. Johnson, A. D., Gomes, W. R., Vandemark, N. L.) New York, London: Academic press 1970

Hoven, H.: Histogenèse du testicule des mammifères. Anat. Anz. 47, 90–101 (1914)

Hösli, P. O.: Zur Problematik der Behandlung des Kryptorchismus. Akt. Urol. 2, 107–120 (1971)

Huggins, C., Moulden, P. V.: Estrogens production by Sertoli cell tumors of the testis. Cancer. Res. 5, 510–514 (1945)

Ito, T., Oinuma, S.: Cytologische Untersuchungen über die Hodenzwischenzellen des Menschen mit besonderer Berücksichtigung auf die Bedeutung von Lipoid und Pigment. Fol. anat. jap. 18, 497–529 (1939)

Jean, C.: Croisance et structure des testicles cryptorchides chez les souris nées de mères traitées a l'oestradiol pendant la gestation. Ann. Endocr. (Paris) 34, 669–687 (1973)

Jean, C. J. M., Jean, Ch., Bergen, M., Turckheim, M. D., Veyssièrs, G.: Estimation of testosterone and androstendione in the plasma and testes of cryptorchid offspring of mice treated with oestradiol during pregnancy. J. Reprod. Fertil. 44, 235–247 (1975)

Jendricke, K., Roger, K., Schäfer, E., Reisert, I.: Vorläufige Ergebnisse klinischer, biochemischer und anatomischer Untersuchungen beim Maldescensus testis. Mschr. Kinderheilk. 121, 634–635 (1973)

Job, C., Garnier, Ph. E., Chaussain, J-L., Toublanc, J. E., Canlorbe, P.: Effect of synthetic luteinizing hormone-releasing hormone on the release of gonadotropins in hypophyso-gonadal disorders of children and adolescents, IV Undescended testes. J. Pediat. 84, 371–374 (1974)

Joachsen, S. G.: Two types of Sertoli cells in man. Acta endocr. 61, 111–116 (1969)

Josso, N.: Müllerian-inhibiting activity of human fetal testicular tissue deprived of germ cells by in vitro irradiations. Pediat. Res. 8, 755–758 (1974)

Mc Kay, D. G., Herzig, A. T., Adams, E. C., Danzinger, L.: Histochemical observations on the germ cells of human embryo. Anat. Rec. 117, 201–219 (1953)

Kessel, R. G.: Annulate lamellae. J. Ultrastruct. Res. Suppl. 10, 1–82 (1968)

Kessel, R. G.: Structure and function of the nuclear envelope and related cytomembranes. In: Progress in Surface and Membrane Science. Vol. 6, p. 243–329. New York: Academic Press, Inc. 1973

Kessel, R. G.: Personal communication (1975)

Kleinteich, B., Schickedanz, H.: Der Spermatogoniengehalt kongenital-dystoper und operativ verlagerter Hoden bei Kindern und Jugendlichen. Z. Urol. 69, 819–824 (1976)

Knorr, D.: Diagnose und Therapie der Descensusstörungen des Hodens. Pädiat. Prax. 9, 229–304 (1970)

Koch, H., Rahlf, G., Kölberling, J., Mühlen, A. V. v. z.: Endokrinologische und morphologische Untersuchungen beim Maldescensus testis. Dtsch. med. Wschr. 100, 682–689 (1975)

Kocher, Th.: Deutsche Chirurgie Lfg. 50 b, 414–532 (1887)

Lahm, W.: Zur Entwicklung der interstitiellen Drüse im Hoden und Ovarium. Mschr. Geburtsh. Gynäk. 58, 128–140 (1922)

Lacy, D., Rotblat, J.: Study of normal and irradiated boundery tissue of the seminiferous tubule of the rat. Exp. Cell. Res. 21, 49–70 (1960)

Lacy, D.: The seminiferous tubule in mammals. Endocrinology 26, 101–105 (1967)

Lacy, D., Lofts, B., Kinson, G., Hopkins, D., Dott, H.: Sertoli cells and steroid synthesis. Gen. comp. Endocr. 5, 693–672 (1968)

O'Leary, J. A.: Comparative studies of the gonad in testicular feminization and cryptorchidism. Fertil. and Steril. 16, 813–819 (1965)

Leeson, C. R., Leeson, T. S.: The postnatal development and differentiation of the boundary tissue of seminiferous tubule of the rat. Anat. Rec. 147, 243–259 (1963)

Leeson, C. R.: An electron microscopic study of cryptorchid and scrotal human testes, with special reference to puberal maturation. Invest. Urol. 3, 498–511 (1966)

Leydig, F.: Zur Anatomie der menschlichen Geschlechtsorgane und Analdrüsen der Säugetiere. Z. wiss. Zool. 2, 1–57 (1850)

Llaurado, J. G., Dominguez, O. V.: Effect of cryptorchidism on testicular enzymes involved in androgen biosynthesis. Endocrinology 72, 780 (1941)

Lubarsch, O.: Ueber das Vorkommen kristallinischer und kristalloider Bildungen in den Zellen des menschlichen Hodens. Arch. Pathol. Anat. Physiol. 145, 316–338 (1896)

Lüdin, A.: Histometrische Untersuchungen an den Spermatogonien bei kryptorchen Knaben. Med. Diss. Basel 1977. In Press.

Mack, W. S.: Infertility and the undescended testicle. Acta endocr. Suppl. 51, 647–651 (1960)

Majno, G., Ryan, G. B., Galbiani, G., Hirschel, B. J., Irle, C., Joris, J.: Inflammation: Mechanism and control (eds. Hess S. M. and Millonig R. C.) p. 13–27. New York, London: Academy Press 1972

Mancini, R. E., Nabraitz, R., Lavieri, J. C.: Origin and development of the germinative epithelium and Sertoli cells in the human testes. Cytological, cytochemical and quantitative study. Anat. Rec. 136, 477–490 (1960)

Mancini, R. E., Vilar, O., Lavieri, J. C., Andrada, J. A., Heinrich, J. J.: Development of Leydig cells in the normal human testis: A cytological, cytochemical and quantitative study. Amer. J. Anat. 112, 203–214 (1963)

Mancini, R. E., Rosemberg, E., Cullen, M., Lavieri, J. C., Vilar, O., Bergada, C., Andrada, J. A.: Cryptorchid and scrotal human testis. I. Cytological, cytochemical and quantitative studies. J. Endocrinol. Metabo. 25, 927–942 (1965)

Mazzuca, M.: Essais d'étude au microscope électronique de biopsies testiculaires fonctionnelles. Lille med. 16, 515–534 (1971)

Michelson, L.: Studies of male fertility: Bilateral lesions of the genital organs, simulating unilateral involvement. Fertil. and Steril. 3, 316–323 (1952)

Monesi, V.: Reproduction in Mammals. In: Germ Cells and Fertilization. (eds. Austin C. R. and Short R. V.) Cambridge Univ. Press 1972

Mott, F. W.: Normal and morbid conditions of the testis from birth to old age in one hundred asylum and hospital cases. Brit. Med. J. 2, 655–698 (1919)

Nagano, Z.: Some observations on the fine structure of the Sertoli cell in the human testis. Z. Zellforsch. 73, 89–106 (1966)

Nelson, W. O.: Effect of gonadotropic hormone injections upon hypophyses and sex accesories of experimental cryptorchid rat. Proc. Soc. exp. Biol. (N. Y.) 31, 1192–1194 (1934)

Nicole, R., Spindler, B.: Prognosis as to fertility following operations for cryptorchidism in children. p. 6. Das Hormon 25, April 1964

Niemi, M., Ikonen, M.: Primordial germ cells in foetal and postnatal human testis. Ann. Med. exp. Fenn. 43, 23–28 (1965)

Numanoglu, I., Köktürk, I., Mutaf, O.: Light and electronmicroscopic examinations of undescended testicles. J. ped. Surg. 4, 614–619 (1969)

Pelliniemi, L., Niemi, M.: Fine structure of the human fetal testis. I. The interstitial tissue. Z. Zellforsch. 99, 507–522 (1969)

Prader, A., Illig, R., Zachmann, M.: Prenatal LH-Deficiency as possible cause of male pseudohermaphroditism, hypospadias, hypogenitalism and cryptorchidism. Ped. Res. 10, 883. (1976)

Raboch, J., Zachor, Z.: Ueber die Fertilität von Männern mit Kryptorchismus. Schweiz. med. Wschr. 85, 1196–1199 (1955)

Raboch, J., Starka, L.: Plasmatic testosterone in bilateral cryptorchids in adult age. Andrologie 4, 107–112 (1972)

Rasmussen, A. T.: Interstitial cells of the testis. Courdry's Spec. Cytol. 3, 1674–1722 (1932)

Raynaud, A.: Modification experimentale de la differention sexuelle des embryons des souris par action des hormones androgenes et oestrogenes. Paris: Herman et Cie 1942

Roosen-Runge, E. C.: Motions on the seminiferous tubules of the rat and the dog. Anat. Rec. 109, 413 (1951)

Roosen-Runge, E. C., Leik, J.: Gonocyte degeneration in the postnatal male rate. Amer. J. Anat. 122, 275–300 (1968)

Rosenmerkel, J. F. (1820) cit. Moormann, J. G.: Histochemische Untersuchungen bei der Lage-anomalie des Hodens nach Behandlung mit humanem Chorion-Gonadotropin (HCG) im Tier-experiment und am Krankengut. Ann. Univ. sarav. Med. 19, 89–144 (1972)

Ross, M. H., Lang, I. R.: Contractile cells in human seminiferous tubules. Science 153, 1271–1273 (1966)

Ross, M. H.: The fine structure and development of the peritubular contractile cell component in the seminiferous tubulus of the mouse. J. Anat. (London) 121, 523–558 (1967)

Ross, M. H., Grant, L.: On the structural integrity of basement membrane. Exp. Cell. Res. 50, 277–281 (1968)

Rothwell, B., Tingari, M. D.: The ultrastructure of boundary tissue of seminiferous tubule in the testis of domestic fowl (Gallus domesticus). J. Anat. (London) 114, 321–328 (1973)

Rowley, M. J., Berlin, J. D., Heller, C. G.: The ultrastructure of four types of human spermato-gonia. Z. Zellforsch. 112, 139–157 (1971)

Russo, J.: The fine structure of the Leydig cell during postnatal differentiation of the mouse testis. Anat. Rec. 170, 343–355 (1971)

Russo, J., Rosas, J. C. de: Differentiation of the Leydig cell of the mouse testis during the fetal period – an ultrastructural study. Amer. J. Anat. 130, 461–480 (1971)

Sapsford, C. S.: The development of the testis of the merino ram, with special reference to the origin of the adult stem cells. Aust. J. Agr. Res. 13, 487–491 (1962)

Schapiro, B.: Ist der Kryptorchismus chirurgisch oder hormonell zu behandeln? Dtsch. med. Wschr. 38, 38–39 (1931)

Schmidt, F. C.: Licht; und elektronenmikroskopische Untersuchungen an menschlichen Hoden und Nebenhoden. Z. Zellforsch. 63. 707–727 (1964)

Schulze, C.: On the morphology of the human Sertoli cell. Cell Tiss. Res. 153, 339–355 (1974)

Scorer, C. G.: The descent of the testis. Arch. Dis. Childh. 39, 605–609 (1964)

Scorer, C. G., Farrington, G. H.: Congenital Deformities of the Testis and Epididymis. London: Butterworth & Co. 1971

Scott, L. S.: Unilateral cryptochidism, subsequent effects on fertility. J. Reprod. Fertil. 2, 54–60 (1961)

Seguchi, H., Hadziselimovic, F.: Morphologie der peritubulären Strukturen des Tubulus seminiferus bei Kindern. Z. mikr.-anat. Forsch. 88, 1149–1160 (1974)

Seguchi, H., Hadziselimovic, F.: Ultramikroskopische Untersuchungen am Tubulus seminiferus bei Kindern von der Geburt bis zur Pubertät. I. Spermatogonienentwicklung. Verh. anat. Ges. (Jena) 68, 133–148 (1974)

Seguy, E.: Ectopic testiculaire et sterilite masculine. Med. Diss. Paris 1961

Sertoli, E.: Dell'esistenza di particolari cellule ramificati nei canalicoli seminiferi del testiculo humano. Morgagni 7, 31–39 (1865)

Setschell, B. P.: Testicular blood supply lymphatic drainage and secretion of fluid. In: The Testis I. (eds. Johnson A. D., Gomes, W. R., Vandemark, N. L.) p. 101. New York, London: Academic Press 1970

Shirai, M., Matsuskita, S., Kagayama, M., Ichijo, S., Takuchi, M.: Histological changes of the scrotal testis in unilateral cryptorchidism. Tohoku J. exp. Med. 90, 363–371 (1966)

Sniffen, R. C.: The testis. I. The normal testis. Arch. Path. 50, 259–284 (1950)

Sohval, A. R.: Histophatology of cryptorchidism. A study based upon the comparative histology of retained and scrotal testes from birth to maturity. Amer. J. Med. 16, 346–362 (1954)

Sohval, A. R., Suzuki, L., Gabrilove, L. J., Churg, J.: Ultrastructure of Crystalloids in Spermato-gonia and Sertoli cells of normal human testis. J. Ultrastruct. Res. 34, 83–102 (1971)

Spangaro, S.: Ueber die histologischen Veränderungen des Hodens, Nebenhodens und Samen-leiters von Geburt an bis zum Greisenalter, mit besonderer Berücksichtigung der Hoden-atrophie, des elastischen Gewebes und Vorkommens von Kristallen im Hoden. Anat. H. 18, 593–771 (1902)

Städtler, F., Moormann, J. G., Hartmann, R.: Morphometrische und histochemische Unter-suchungen an kindlichen Hoden unter besonderer Berücksichtigung des Kryptorchismus. Saar. Aerztebl. 23, 430–434 (1970)

Städtler, F., Hartmann, R.: Histologische und morhpmetrische Untersuchungen zum präpuberalen Hodenwachstum bei normal entwickelten und cerebral geschädigten Knaben. Dtsch. med. Wschr. 97, 104–109 (1972)

Städtler, F.: Die normale und gestörte präpuberale Hodenentwicklung des Menschen. Stuttgart: Gustav Fischer 1973

Stieve, H.: Männliche Genitalorgane. In: Handbuch der mikroskopischen Anatomie des Menschen. Bd. 7 (ed. von Möllendorff) S. 40–51, Berlin: Springer 1930

Stieve, H.: Die Entwicklung der Keimzellen und der Zwischenzellen in der Hodenanlage des Menschen. Z. mikr.-anat. Forsch. 10, 225–285 (1928)

Usadel, K. H., Leuschner, U., Schwedes, U., Schade, C., Schöffling, K.: Ueber das Auftreten cholinerger Nervenfasern in den Tubuluswänden des Hodens bei primären Hypogonadismus (Klinefelter-Syndrom, Kryptorchismus). Klin. Wschr. 51, 1016–1023 (1973)

Vilar, O.: Histology of the human testis from neonatal to adolescence. The Human Testis. Plenum Press New York (eds. Rosenberg E. and Paulsen C. A.) p. 95–108 (1970)

Von La Valette St. George: Ueber die Genese der Samenkörper. Arch. mikr. Anat. 15, 261 (1876)

Wartenberg, H., Holstein, A. F., Vossmeyer, J.: Zur Cytologie der pränatalen Gonadenentwicklung beim Menschen. II Elektronenmikroskopische Untersuchungen über die Cytogenese von Gonocyten und fetalen Spermatogonien im Hoden. Z. Anat. Entwickl.-Gesch. 134, 165–185 (1971)

Yusawa, J.: An Elektron Microskopic Study of the Boundary Tissue of the Human Seminiferous Tubule. Nippon Hiniokika Gakkai Zasshi Tokyo 59, 294 (1968)

Subject Index